Plant Biology
Laboratory Manual
Almuth Tschunko
Marietta College

GRAHAM · GRAHAM · WILCOX

PEARSON

Prentice
Hall

Upper Saddle River, NJ 07458

Editor-in-Chief: Dan Kaveney
Senior Acquisitions Editor: Andrew Gilfillan
Project Manager: Crissy Dudonis
Illustration and Text Design: Vanessa Handley
Executive Managing Editor: Kathleen Schiaparelli
Senior Managing Editor: Nicole M. Jackson
Assistant Managing Editor: Karen Bosch Petrov
Production Editor: Traci Douglas
Supplement Cover Manager: Paul Gourhan
Supplement Cover Designer: Christopher Kossa
Manufacturing Buyer: Ilene Kahn
Manufacturing Manager: Alexis Heydt-Long

© 2007 Pearson Education, Inc.
Pearson Prentice Hall
Pearson Education, Inc.
Upper Saddle River, NJ 07458

Printed in the United States of America

10 9 8 7 6 5 4 3 2 1

ISBN 0-13-143508-6

Pearson Education Ltd., *London*
Pearson Education Australia Pty. Ltd., *Sydney*
Pearson Education Singapore, Pte. Ltd.
Pearson Education North Asia Ltd., *Hong Kong*
Pearson Education Canada, Inc., *Toronto*
Pearson Educación de Mexico, S.A. de C.V.
Pearson Education—Japan, *Tokyo*
Pearson Education Malaysia, Pte. Ltd.

This lab manual is dedicated to
the memory of

"Dr. Nick"
Dr. Norton H. Nickerson
(1926-1999)

inspiring professor of
botany and environmental science
at
Tufts University
(1963 to 1996)

PREFACE

This Plant Biology Laboratory Manual uses an *investigative approach*, actively engaging the student. The investigations have been student tested to various degrees, some for many years.

The goal of this manual is to enable students to acquire a more holistic understanding of plants—what plants do during the day and night, through the seasons, and through the years, as well as what their roles are in the ecosystems. The topics are always related to applications in everyday life. All too often, students get so consumed with trying to master terminology and molecular details that they "fail to see the forest for the trees."

This lab manual is designed to include the following features:
- An *investigative* approach; the investigations range from being partly to wholly student designed.
- *Themes of ecology, evolution, and genetics* are integrated throughout, as in the *Plant Biology* textbook by Graham, Graham, and Wilcox.
- *Versatility; choices* within labs make the manual usable in a *wide range of situations*—
  - in all parts of the country, at all seasons.
  - in labs consisting of two- or three-hour sessions.
  - in variously designed plant biology courses.
    - Labs are self-contained and can be done in any order (the manual can thus be a companion to the Graham, Graham, and Wilcox textbook, *Plant Biology*, or any other botany textbook).
    - The investigations range from hypothesis testing, to observation, to information retrieval and discussion.
    - The investigations are of various durations, from one lab period to a few weeks to the entire term.
- *Numerous outdoor investigations*; these focus on a *variety of topics*, not just plant identification. There's nothing like studying the living plants outdoors. (In my view, studying botany from specimens on styrofoam trays or from pressed specimens is a last resort.)
- *Minimal budget* is required; the labs use basic lab equipment and materials.
- *Complete lists of materials* are provided for the instructor's and the student's convenience:
  - *Catalog numbers of two large suppliers* (Carolina Biological Supply Company and Ward's) are provided for specialty items, although these items are available from other sources as well. If Carolina and Ward's do not currently carry the items, the catalog numbers of other suppliers are provided. The living plant materials may also be available in your campus greenhouse, on your campus grounds, or in nearby areas.
  - Solutions and setup are fully described in the lab, rather than in a separate instructor's manual. This approach fully informs the students, as well as allows students who later find themselves in a teaching situation to use their old lab manual as a reference—without facing the impossible task of tracking down an instructor's manual for the full details.
- *Minimal terminology* is used, although there is enough to ensure comprehension and communication. The student's textbook provides further details. When new terms are introduced, the meanings of their *Greek*

and Latin roots are provided in order to help the student learn the terms by understanding them rather than by blindly memorizing them (Latin: *memor* = mindful).

- *Appendices* assist students with *numerical analysis of data*—appendices that cover the metric system, data presentation (tables and graphs), and statistics (mean and standard deviation; t-test and chi-square test).
- *Quick orientation features* in the table of contents make it a better resource. Each lab's descriptive title is followed by subheadings, with quick-reference symbols included to indicate the type of investigation:

✪ entirely student designed	📖 information retrieval	🗣 discussion
🕐 long-term experiment	✎ numerical analysis	◉ microscopy involved
✿ done outdoors	☐ could be done partly or entirely outdoors	

Also, the duration of an investigation is indicated if it is long-term.

The labs are organized to include these features:
- Objectives
- Background, concise but sufficient
- Materials, described fully
- Investigation, described in steps
    - Procedures and/or thoughts to consider when designing the investigation are provided.
    - Check-off box precedes and highlights each numbered step.
- Data sheets and worksheets
- Questions throughout the lab require students to analyze the investigation critically and to relate the results to the bigger picture.
- Suggested further investigations or extended assignments, which the instructor may choose to assign
- Literature cited section, if print references, videos, and/or websites were cited in the lab

Experimental design. Numerous investigations in this lab manual require the students to design the experiment, either partially or entirely. To prepare students for this, it is recommended that Lab 1 be done in the first lab meeting in order to set the stage. A bonus is that Lab 1 serves as a great icebreaker in the first meeting.

Choices. Due to each lab having several parts, the instructor can choose the investigation(s) that best fit(s) the course. If the instructor wants to use several parts but time is limited, the instructor may have different student groups do different parts and then orally report their results to the class.

A note about websites. Students gravitate to websites. This lab manual mentions various print references as well as several carefully selected websites having useful and unique information pertaining to the labs. Students need guidance in the careful use of the web—in recognizing quality websites as well as in avoiding those likely to contain inaccurate information. It is important for students to realize the shortcomings of websites: unlike peer-reviewed print references, the websites frequently have no quality control, and the contents as well as addresses can change at any time. If a web address has changed since the manual went to press, students should do an Internet search for the web site's title or sponsor. (Some instructors choose to ban websites as references in the student lab reports, requiring students to consult print references and primary literature.)

Suggestions? If you have any suggestions for improving this manual, or if you spot any errors that have escaped my notice, your comments would be most welcome and appreciated. You may e-mail me at almuth.tschunko@marietta.edu.

Acknowledgments

Writing and publishing a lab manual involves the efforts of many people. I am grateful to Linda Graham, lead author of the *Plant Biology* textbook, for suggesting my name to Prentice Hall and for numerous productive discussions. I truly appreciate the contributions of the talented people at Prentice Hall who have shepherded this project to completion: Travis Moses-Westphal (project manager) initiated the project, Colleen Lee (assistant editor) managed the first half of the work, Karen Horton (project manager) managed the second half of the work, and Crissy Dudonis (project manager) brought the project to completion. Also at Prentice Hall, sincere thanks to Sheri Snavely (editor at the start of this project) and Andrew Gilfillan (editor at the conclusion of this project), Karen Bosch (managing editor), and Traci Douglas (production editor). In addition, I wish to acknowledge the careful work of the copy editor Christianne Thillen and the proofreader Colleen Lee Moore. It has also been my good fortune to work closely with Vanessa Handley, a talented illustrator and layout person (with the added bonus of having a PhD in plant biology), to make the material most easily accessible to the student.

My thanks to the following people for kindly granting permission to use the chromosome squash procedure in the mitosis lab: Joseph J. Nickolas (now Professor Emeritus, Northland Pioneer College, who first presented this procedure in a workshop at the National Association of Biology Teachers (NABT) Convention, University of Nevada, Las Vegas, on October 24, 1981), Connie Roderick (who made some modifications while at the University of Wisconsin–Baraboo/Sauk County; she is presently at USGS, National Wildlife Health Center, Madison, Wisconsin), and Glen G. Wurst (Allegheny College, who made some further modifications).

Pearson Education, Inc. (Upper Saddle River, New Jersey) granted permission to reprint, in Appendix 3 of this lab manual, Tables B-1 and B-3 of Appendix B from the book by Jerrold H. Zar, *Biostatistical Analysis*, 4th ed., © 1999.

It is a privilege to thank my students of two decades of teaching; I continue to learn from all of them. Thanks in particular for the help provided by these students: Rebecca Stewart taught me how to take digital photos of microscope slides for this manual, and Elizabeth Pierson, Laura Fitzsimmons, Racheal Desmone, and Sarah Lane tried out some ideas for this manual.

I am grateful to my colleagues in the biology department at Marietta College for the continued intellectual stimulation they provide. Thanks especially to David McShaffrey and David Brown for cheerfully helping me for years when I have been stumped with computer and software problems.

References consulted for the etymology of scientific terms:
Webster's Collegiate Dictionary, 11th ed. 2003. Springfield, MA: Merriam-Webster, Inc.
Jaeger, Edmund C. 1955. *A Source-book of Biological Names and Terms*, 3rd ed. (fourth printing). Springfield, IL: Charles C. Thomas.

Reviewer Acknowledgments

My thanks and appreciation go to the following reviewers who read various portions of this lab manual, providing very helpful feedback:

Teresa M. Barta, *University of Wisconsin–Stevens Point*
Richard R. Bounds, *Mount Olive College*
Jill Bushakra, *Cabrillo College*
Roger del Moral, *University of Washington*
Kathy Falkenstein, *Hood College*
Joseph Faryniarz, *Naugatuck Valley Community College*
Nabarun Ghosh, *West Texas A&M University*
Linda Graham, *University of Wisconsin–Madison*
Everett Hansen, *Oregon State University*
Pamela Hellman, *Texas Tech University*
Carolyn J. P. Jensen, *Pennsylvania State University*
William A. Jensen, *The Ohio State University*
Tasneem Khaleel, *Montana State University–Billings*
David W. Kramer, *The Ohio State University–Mansfield*

Edgar Moctezuma, *University of Maryland*
John Olsen, *Rhodes College*
H. Carol Reiss, *Brown University*
Susanne Renner, *University of Missouri–St. Louis*
Magaly Rincón-Zachary, *Midwestern State University*
Bruce Serlin, *DePauw University*
Brian Shmaefsky, *Kingwood College*
Teresa Snyder-Leiby, *State University of New York–New Paltz*
F. Lee St. John, *The Ohio State University–Newark*
Robb D. VanPutte, *McKendree College*
Thad E. Yorks, *California University of Pennsylvania*
Carol L. Wymer, *Morehead State University*

Photo and Illustration Credits

This lab manual is enhanced by the excellent photos and illustrations. The following figures of this lab manual were from Lee Wilcox and appear in the Graham, Graham, and Wilcox textbook *Plant Biology*, 2nd edition (the asterisks indicate Wilcox figures that illustrator Vanessa Handley modified to suit the needs of this lab manual): Figures 3.1*, 3.2*, 4.1*, 5.3*, 5.4, 6.1a*b*, 7.2*, 9.2, 9.4*, 9.5*, 10.2 (all except the potato), 10.3*, 10.4 (the monocot and the dicot roots and stems), 11.1*, 11.3abc, 11.4ab, 11.5 (all photos), 14.1*, 15.1abc, 15.2 (all photos except *Spirogyra* and *Chondrus*), 15.3ab, 15.4 (all ten photos), 17.1 (the photo), 17.2, 17.3*, 17.4*, 17.5a*, 17.7, 18.1*, 18.2*. David Brown photographed Figure 18.7; Martha Cook photographed *Spirogyra* in 15.2; the author photographed the microscope slides of *Tilia* root and stem of 10.4 and the pressed *Chondrus* of 15.2 (from Carolina Biological Supply Company), and created 11.2. Vanessa Handley created all of the remaining illustrations in this lab manual.

TABLE OF CONTENTS

✪ entirely student designed　　　📖 information retrieval　　　🗣 discussion
🕐 long-term experiment　　　✏ numerical analysis　　　⊙ microscopy involved
☼ done outdoors　　　⬚ could be done partly or entirely outdoors

✪ entirely student designed 📖 information retrieval 🗣 discussion
⏱ long-term experiment ✎ numerical analysis ⊙ microscopy involved
✿ done outdoors ▯ could be done partly or entirely outdoors

✪ entirely student designed	📖 information retrieval	🗣 discussion
◷ long-term experiment	✎ numerical analysis	◉ microscopy involved
☼ done outdoors	⬚ could be done partly or entirely outdoors	

✪ entirely student designed
⏱ long-term experiment
✿ done outdoors

📖 information retrieval
✐ numerical analysis
⬚ could be done partly or entirely outdoors

✊ discussion
⊙ microscopy involved

✪ entirely student designed	📖 information retrieval	🎙 discussion
🕐 long-term experiment	✐ numerical analysis	⊙ microscopy involved
☼ done outdoors	⬚ could be done partly or entirely outdoors	

1

DESIGN A RESEARCH PROJECT (discussion only)

Science is, I believe, nothing but trained and organised common sense.

—T. H. Huxley
(1825–1895)
On the Educational Value of the Natural History Sciences
(1854)

1. Learn about the scientific method.
2. Learn the characteristics of proper experimental design.
3. Learn about poison ivy.
4. Gain experience in information retrieval.

BACKGROUND: POISON IVY RASHES—JEWELWEED TO THE RESCUE?

Figure 1.1 shows poison ivy and poison oak. Some people are allergic to these plants and get an itchy red rash with blisters after touching the plants. Other people are not allergic, although they may become so during their lifetimes. Identifying these plants is challenging due to the variability in leaves and in growth habit. To make matters worse, allergic people can get a rash not only directly from touching any part of the plant (even in winter) but also indirectly by petting an animal or touching clothing that has brushed against these plants and thus picked up their oils.

You have a pretty good chance of running into these plants since all of the lower 48 states of the United States as well as southern Canada have one or another of these species. The eastern poison ivy (scientific name *Toxicodendron radicans*) occurs throughout the East; the western poison ivy (*Toxicodendron rydbergii*) is found throughout the West although it also extends eastwardly into the northern United States and adjacent Canada; Atlantic poison-oak (*Toxicodendron pubescens*) grows in sandy woods of the southeastern United States; Pacific poison-oak (*Toxicodendron diversilobum*)

is in the far western states. Earlier botanists had classified these species in various ways; for example, the name *Toxicodendron radicans* at one time included not just the eastern poison ivy but also the western poison ivy and the Pacific poison-oak. Note that the first word of the scientific name is the genus; closely related species have the same genus name.

Figure 1.2 shows jewelweed, also known as touch-me-not or by its scientific genus name *Impatiens*. Its fruit consists of a pod containing two seeds typically. A mature pod rips at the slightest touch into five coiled strips, explosively expelling the seeds—hence the names of touch-me-not and *Impatiens*. Two species, the spotted jewelweed (*Impatiens capensis*) with orange flowers and the pale jewelweed (*Impatiens pallida*) with yellow flowers, together cover the area east of the Rockies in the United States and adjacent Canada, while *I. capensis* also occurs in the Pacific Northwest. These plants grow in moist soil such as along streams and in ditches.

A widely known folk remedy thought to promote the healing of a poison ivy rash (and other rashes) is to apply the juice of crushed stems or leaves of jewelweed to the rash. This remedy was used by Native Americans and settlers alike and has extended to the present day. The garden *Impatiens* has also now been used for this purpose.

The question is: does jewelweed juice really speed the healing of a poison ivy rash?

eastern poison ivy

Figure 1.2 Spotted jewelweed

Pacific poison-oak

Figure 1.1 Poison ivy, poison-oak

INVESTIGATIONS

(No advance preparation by the student is needed for this discussion.)
In class:

☐1. If poison ivy or poison-oak and jewelweed occur in your immediate area, your instructor may take you outside to see them and learn how to identify them. Otherwise you can see color images of these species on The PLANTS Database website of the U.S. Department of Agriculture, at http://plants.usda.gov.
NOTE: Website addresses may change; if you cannot locate a website by a given address, then do a web search for the organization and/or the name of the database.)

☐2. The class will be divided into groups of about five students, by counting off.

☐3. Each group will represent a "research group" at a pharmaceutical company given the assignment of designing a research project to test the hypothesis that jewelweed juice speeds the healing of a poison ivy rash.

☐4. You will be given a set amount of time (about 20 minutes) for group discussion to plan out the research project in detail. *You will not actually conduct this research*. How many subjects would be used, and how would they be treated? What *quantitative* data would be collected, and how would they be collected? (Also, how would you get subjects to take part in this study?)

☐5. After the allotted time, a spokesperson from each student group will orally present the group's experimental design.

DO NOT READ BEYOND THIS POINT until all groups have presented their experimental design.

☐6. After the group presentations are finished, the instructor will guide the class as a whole through a discussion of various points such as:

• What are the good features of each group's design, and why are these features important?

• What are the pros and cons of *in vivo* vs. *in vitro* experiments (Latin: *vivus* = alive, *vitrum* = glass)? What are the pros and cons of various animal species used as subjects for *in vivo* experiments?

• What is the importance of **control** group vs. **experimental** groups?

• What is the importance of **replication**? How many subjects should there be in each group?

• To ensure that there is **only one variable** in the experiment, how would you design all phases of the experiment—from selection of subjects, to application method for the lotions, to evaluation of the rash status of the subjects in the groups?

• What is a **placebo**, and why is it important (especially with human subjects)?

• Why is it important to have a **double-blind experiment** (neither the subject nor the evaluator knows who got what treatment)? How would this be accomplished?

• What are the pros and cons of applying the placebo and the jewelweed lotion on the *same* subject vs. on *different* subjects?

• What about the possibility of a subject being allergic to the placebo or to the jewelweed lotion? Can the experiment be designed to detect this?

• How do you **quantitate** the "healing rate of a rash"? Which would be easiest to quantitate: itchiness, redness, or blister status? How would you quantitate it?

• How are you going to **recruit subjects** for this experiment? Who would volunteer to be given a rash? What would it take to get you to be a subject? What is your price? $50? $100?

• OK, how much money would then be required to pay all the subjects needed for this research? Where is this money going to come from? This leads into a discussion on **funding** sources for medical research (NSF, NIH, etc.), **grant writing**, grant review process and funding priority, and the time that elapses from the initial idea to the actual start of a research project.

• What **ethical consideration**s and institutional policies and/or requirements are there when using human subjects in research? What is in an "Informed Consent Form"?

☐7. The discussion will conclude with the class formulating an improved experimental design, then critiquing it and revising it repeatedly until it looks solid (follows proper scientific method and will produce reliable data).

NOTE: Researchers follow a very similar process when they plan out their research. Scientists with a research question in mind will first review the scientific literature in that area. Then they brainstorm, evaluate their ideas, and come up with an experimental design, which they revise until it looks good. Then they often discuss the design with colleagues to help locate hidden flaws and trigger ideas for further improvements in the design. The experimental design can go through several rounds of revision before the experiment is actually performed. Designing and planning out a good experiment may well take longer than actually performing the experiment; time invested in planning is time well spent, since it saves effort and money in the long run.

EXTENDED ASSIGNMENTS

1. **Research the various plant species mentioned in this lab.** For each species, what are the distinguishing characteristics, habitat, and range? To which plant family does it belong, and what are some other members of this family?

Start by consulting The PLANTS Database website of the U.S. Department of Agriculture: http://plants.usda.gov. On this extensive website, you can do a plant *name search* at the top of the homepage (in one box select either *scientific name* or *common name*, and in the other box type the name to be searched). If you search for a species, you will be brought to its page, which has a "plants profile" that includes its distribution map for the United States. If you do a search for the genus name *Toxicodendron*, the list of entries that appears contains an entry that is just the genus *Toxicodendron* itself as well as entries of the scientific names of the various species and subspecies within this genus; by clicking on the list's entry that is the genus itself, you will get to the page for this genus, which includes a distribution map of the United States for this genus, followed by an individual U.S. distribution map *for each species in this genus*. These individual maps give you a very quick, convenient overview of the distribution of the various species of *Toxicodendron* in the United States. For many of the states on a range map, you can click and find a detailed map of that state showing the counties in which the plant occurs.

On the Botanical Society of America website at http://www.botany.org/, go to "Suggested WWW Links" and click on "General Links"; explore the listed major botany links and resources on the web. Select a few sites and see what you can find there on these plants. (Use a newer browser to see all the menu bars.)

NOTE: The scientific names have changed over the decades as botanists have reclassified the plants. Eastern poison ivy *Toxicodendron radicans* had previously been called *Rhus radicans* and *Rhus toxicodendron*; Atlantic poison-oak *Toxicodendron pubescens* has had various other names previously, among them being *Toxicodendron toxicaria*, *Rhus quercifolia*, and *Rhus toxicodendron* (Gleason and Cronquist 1991, Fernald 1950). See also the PLANTS Database website for additional synonyms.

2. **Can you find any published sources giving solid experimental evidence for the effectiveness of *Impatiens* against *Toxicodendron* rashes?** (The labor of doing this search can be divided among the student groups.) Note that the U.S. Food and Drug Administration (FDA) considers herbal medicines to be "dietary supplements" rather than drugs, and thus they are not subject to the same rigorous clinical trials and manufacturing standards as are FDA-approved drugs.

Do a search to find literature on the use of jewelweed to heal poison ivy rashes. Use several different electronic databases such as Biosis and MedLine. However, note the *years covered by the various databases*. If you restrict your search to electronic databases only, you will miss literature that was published in earlier years. Because the use of jewelweed as a folk remedy has a long history, you may find articles in the literature from decades ago. To get to the earlier articles, you can use print indexes like BioAbstracts. There is also an electronic database that can be searched for older literature; called JSTOR, it is an archive of certain scholarly journals; you can search JSTOR at http://www.jstor.org if your institution has a subscription. (Note that older journal articles may refer to poison ivy by its previous genus name, *Rhus*.)

Do an Internet search for websites that mention jewelweed and poison ivy. Can you find any published scientific research cited, or are there only anecdotal claims and testimonials of the efficacy of jewelweed juice? Remember that anyone can claim anything on a website because the web requires no peer review for quality control. In contrast, articles in reputable scientific journals are peer reviewed; this means that a submitted manuscript must be reviewed by experts in that field and be recommended for publication.

Evaluate all of the sources that you have found to determine whether they are primary literature (presenting new experimental data), or secondary litera-

ture (review articles, with sources cited), or tertiary literature (for the general public, with no citations although "further readings" may be listed at the end).

Write a brief report or give a brief oral presentation describing the research from any primary literature source that you have located.

LITERATURE CITED

Botanical Society of America. 2006. Botanical Society of America "Additional Botanical Links," http://www.botany.org/newsite/education/wwwlinks.php (accessed February 19, 2006).

Fernald, M. L. 1950. *Gray's Manual of Botany*. 8th ed. New York: American Book Company.

Gleason, H. A., and A. Cronquist. 1991. *Manual of Vascular Plants of Northeastern United States and Adjacent Canada*. 2nd ed. Bronx, NY: New York Botanical Garden.

Huxley, T. H. 1968. *Science and education essays: On the educational value of the natural history sciences* (1854). New York: Greenwood Press.

Journal Storage (JSTOR). 2006. Journal Storage, http://www.jstor.org (accessed February 19, 2006).

USDA, Natural Resources Conservation Service. 2006. The PLANTS Database, Version 3.5, http://plants.usda.gov (accessed February 19, 2006).

LAB 2

MICROSCOPY AND PLANT CELLS

I.   Getting to know your compound microscope
II. Plant cells are dynamic!
III. Diversity in plant cells
IV. Osmosis: effect of road salts

BACKGROUND

In forensic labs, microscopists who are well trained in plant biology can use the unique microscopic features of plant cells to identify a small plant fragment. These unique microscopic features help to solve crimes by matching fragments from the crime scene and fragments on a suspect's clothing, for example. Likewise, microscopic analysis of the stomach contents of a corpse will reveal what the individual had last eaten, which may be helpful when investigating the individual's death. There is a real need for plant scientists who are willing to become involved in forensic investigations and who are well trained in plant anatomy, taxonomy, DNA analysis, ecology, and palynology (the study of pollen and spores).

In this lab, you will first learn to use a microscope properly; then you will learn about plant cells. In textbooks, a diagram of a general plant cell typically looks like a brick. In this lab, you will observe that plant cells come in various shapes and sizes and have various inclusions; all of these characteristics help in identifying forensic plant specimens. You will gain further insight into the diversity of plant microscopic structures in Labs 9 (Tree Investigations), 10 (Roots), and 11 (Leaves).

PART I:
GETTING TO KNOW YOUR COMPOUND MICROSCOPE

OBJECTIVES

1. Learn how to carry a microscope properly.
2. Learn the parts of a microscope so that you can converse effectively about it.
3. Learn how to operate a microscope properly and focus on a specimen using each objective lens.
4. Practice moving a specimen in the desired direction under a microscope and focusing up and down through it.

BACKGROUND

The parts of the compound microscope are shown in Figure 2.1. The name "**compound** microscope" indicates that the microscope contains two main lens systems. Light passes through the specimen and then through the first lens system (includes the objective), which magnifies the specimen and forms an image within the body of the microscope. This image is then further magnified by the second lens system (includes the eyepiece).

Microscopists use the terms magnification, resolution, and contrast. **Magnification** is the apparent increase in size of an image. **Resolution** is the ability to distinguish two closely spaced points as being two points rather than one. **Contrast** is the difference in brightness between a dark structure and its lighter surroundings.

Carefully read and follow the instructions for using a microscope. A compound microscope is a very expensive instrument. Improper use can result in a major repair bill. *Ask if you are in doubt!*

Rules for Proper Handling of a Microscope

1. *Always carry a microscope upright, with **two hands***—one hand holding the arm of the microscope and the other hand supporting the base of the microscope.
2. *Never remove the eyepiece.* Removing the eyepiece can put its lenses out of alignment and admits dust onto the internal optical surfaces; both of these situations require repairs.
3. *Never remove an objective.* This again admits dust and creates the risk of dropping the expensive objective.
4. *Never use force to move a microscope part.* All microscope parts that are intended to move, will move easily. If they do not, then notify the instructor.
5. *Never touch lenses with your fingers.* Fingerprints are bad for the lens coating and blur the image.
6. *Use only lens paper to clean lenses.* Other tissue paper will scratch lenses.

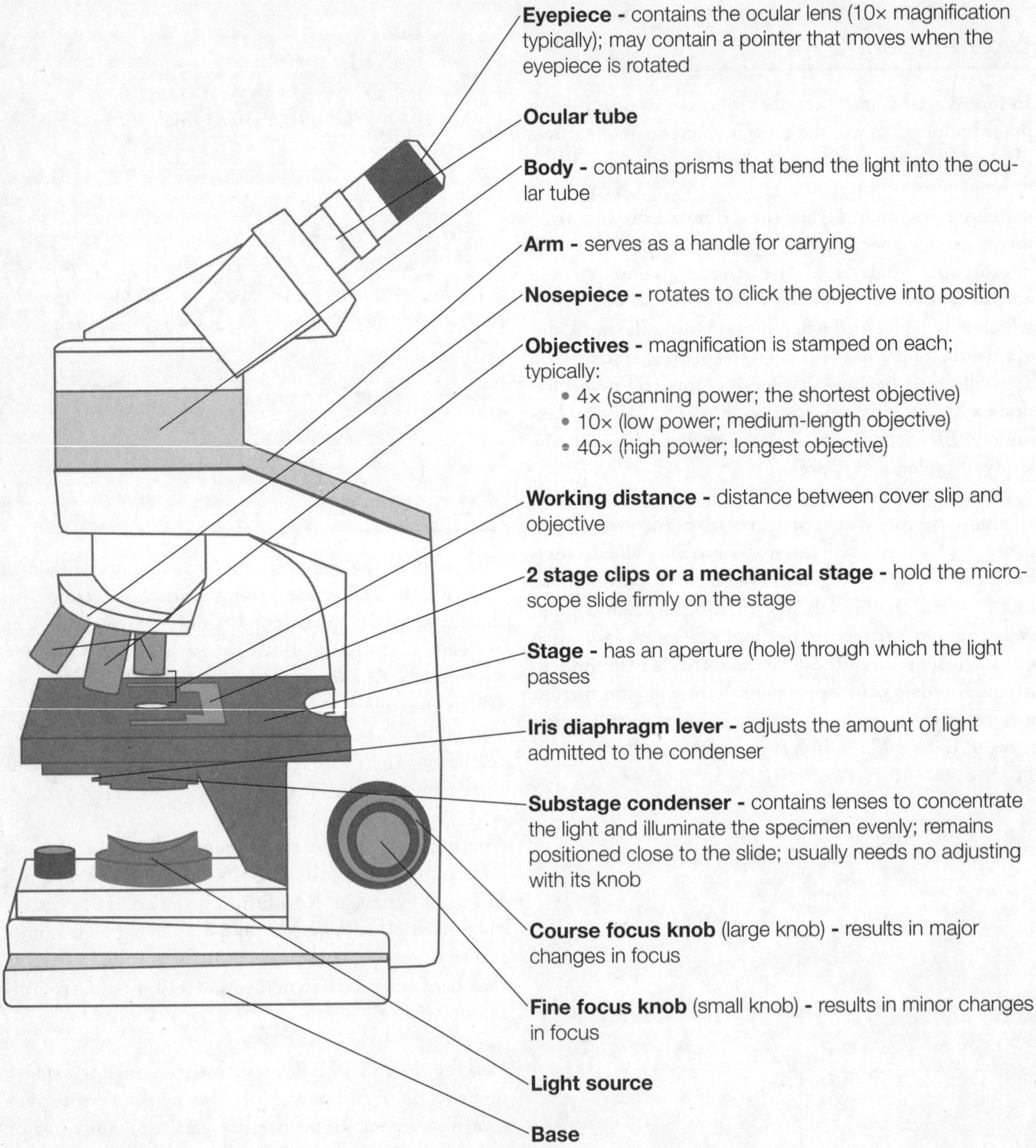
Eyepiece - contains the ocular lens (10× magnification typically); may contain a pointer that moves when the eyepiece is rotated
Ocular tube
Body - contains prisms that bend the light into the ocular tube
Arm - serves as a handle for carrying
Nosepiece - rotates to click the objective into position
Objectives - magnification is stamped on each; typically:
• 4× (scanning power; the shortest objective)
• 10× (low power; medium-length objective)
• 40× (high power; longest objective)
Working distance - distance between cover slip and objective
2 stage clips or a mechanical stage - hold the microscope slide firmly on the stage
Stage - has an aperture (hole) through which the light passes
Iris diaphragm lever - adjusts the amount of light admitted to the condenser
Substage condenser - contains lenses to concentrate the light and illuminate the specimen evenly; remains positioned close to the slide; usually needs no adjusting with its knob
Course focus knob (large knob) **-** results in major changes in focus
Fine focus knob (small knob) **-** results in minor changes in focus
Light source
Base

Figure 2.1 Parts of a compound microscope

PROCEDURE FOR USING A MICROSCOPE

□1. Plug in the microscope cord.

□2. Turn on the light; if there is a dimmer switch, turn it to 2/3 or 3/4 brightness.

□3. Prepare the microscope slide:

a. If it is a prepared **permanent mount**, wipe both sides of it with lens paper to remove all dust and fingerprints. *One fingerprint will make everything look fuzzy!*

b. If it is a **wet mount**, get a microscope slide and cover slip; wipe them clean with lens paper if you detect dust and fingerprints.

- Always place your specimen in a drop or two of water on the microscope slide.
- Always add a cover slip. To avoid air bubbles, touch one edge of the cover slip to the microscope slide and lower the slip onto the water drop.

An air bubble, when in focus, looks like a smooth, black bicycle tire. Air bubbles can actually be very useful—get the edge of an air bubble into crisp focus, and the mounted specimen will also be in focus.

c. Water should be present under the entire cover slip. If there is not enough or barely enough water under the cover slip, add a drop to the edge of the slip—the drop will flow under the slip.

d. If there is too much water, absorb the excess with a tissue or bit of paper towel.

e. *There should be no water on top of the cover slip!*

□4. Rotate the nosepiece and click the **scanning objective** (the shortest objective) into position. *Always start with the short scanning objective.*

□5. Only now, place the prepared slide under the stage clips; position the slide so that the specimen is in the center of the circular aperture (hole) of the stage.

□6. Make sure that the substage condenser is slightly less than fully raised. No further adjustments of the condenser are needed during the class period.

□7. If your microscope has only one eyepiece, you may see the image better if you cup your hand and loosely cover your unused but open eye with it. If your microscope has two eyepieces, look through it with both eyes and adjust the distance between the eyepieces to fit the interpupillary distance of your eyes (your instructor will show you how to adjust this).

□8. Use the iris diaphragm lever to adjust the amount of light and contrast. If there is too much light, it is like looking into a spotlight—the contrast is poor and you will see very little.

□9. Move the slide until you see part of the specimen or an air bubble (usually rather blurry); use the coarse focus knob to get approximate focus.

□10. Use the fine focus knob to get precise focus; continue using the fine focus knob to focus up and down through the specimen.

TROUBLESHOOTING

Your Problem	Try These Actions
The circle of light looks odd, not crisp.	• Make sure that the objective is clicked into position.

You cannot find the specimen.	• Center it in the aperture of the stage.
	• Make sure that you are in the proper focal plane:
	  – If an air bubble is present, center it and get its edge in focus.
	  – Push down on a side of the cover slip; specks in the water should move.
	• Search systematically back and forth across the entire cover slip.

You cannot get a clear image.	• Clean the slide, eyepiece, and objective lens with lens paper. • Adjust the iris diaphragm. • Move the condenser down just a little if it is at its maximum height.
You reach the fine focus knob's stop before getting fully focused.	• Return to the previous objective; back up a few turns on the fine focus, refocus with the coarse focus, and refocus with the fine focus; then return to your objective and fine focus again.
Weird stringy things are floating across the image.	• If the stringy things move when you move your eye, they are "floaters" in the aqueous humor of your eye; that's normal.

11. When you are ready to view a part of the specimen with a higher power objective, first *center* that part of the specimen, and then rotate the nosepiece to click the *next higher power objective* into position. (Modern microscopes are designed to be **parfocal**—when in focus with the scanning objective, they will also be in approximate focus with the other objectives.)

12. Again, use the iris diaphragm lever to adjust the brightness and contrast of the image under this new objective.

13. Adjust only the fine focus knob to focus the image under this new objective.

14. When you are ready to view a particular feature of the specimen with the long, high-power objective, first center the feature in your field of view. If your slide has a thick specimen or cover slip, there may not be enough room for the high power objective to be rotated into position. Therefore, **watch from the side** as you gently rotate the high-power objective into position, to make sure that the objective does not touch the slide.

NOTE: Never force the high-power objective into position. If there is any resistance, stop so that you do not break the cover slip and scratch the objective's lens—which is very expensive to replace; instead, go back to the previous objective to view that slide.

15. Adjust the iris diaphragm lever and refocus with only the *fine focus knob*.

16. When you are describing the position of something in the field of view, refer to the clock face: "It's located at 3 o'clock, halfway to the center," for example.

17. *Always return to the scanning objective before removing the microscope slide from the stage!* This avoids scratching the objective lens because the short scanning objective provides the most working distance. **Working distance** is the distance between the microscope slide's cover slip and the lens at the end of the objective. High-power objectives have a working distance of only about 1 mm.

18. When you are finished using your microscope and have removed the last slide as in step 17, make sure that the stage is clean and dry; turn down the dimmer switch and turn off the light; wrap the power cord around the microscope and tuck in the plug; cover with the plastic dust cover and return the microscope to its assigned spot in the cabinet.

MATERIALS

* microscope materials:
  - compound microscope, 1 per student
  - per lab table: lens paper
                    microscope slides, cover slips
                    dropper bottle with water
* *Elodea*—living

INVESTIGATIONS

1. Study the parts of the microscope and their functions, as shown earlier in Figure 2.1. Find all the parts on your microscope. *Do not remove the eyepiece and objectives.* Your instructor will point out any differences between the diagram and your microscope. Note that the various brands and models of microscopes differ in:
   * location of the coarse and the fine focus knobs.
   * number of objectives and their power.
   * location of the power switch for the light.
   * whether there is also a dimmer switch for the light.
   * whether it is the stage or the objective that moves vertically when focusing.
   * whether the slide is secured to the stage by two stage clips or by a mechanical stage.
   * whether there are one or two eyepieces.

	Magnification Produced by That Part	Total Magnification of Specimen	Cover slip's width (fields of view)
Scanning power objective			
Low-power objective			
High-power objective			
Ocular lens in eyepiece			

Table 2.1 Characteristics of your microscope

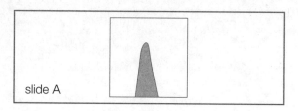

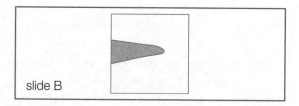

Figure 2.2 Slides A and B of *Elodea*

☐2. In Table 2.1, record the magnification of your objectives (such as "4×") and your ocular lens (usually 10× if it is not stamped on the eyepiece); calculate and record the total magnification of an object viewed under each objective.

total magnification = objective's magnification multiplied
by the eyepiece's magnification

(Note that the last column in Table 2.1 will be filled in later.)

☐3. Prepare two wet mounts of *Elodea* leaves (see step 3b in "Procedure for Using a Microscope"). On one slide, have the leaf tip pointing up (slide A); on the other slide, have the leaf tip pointing to the right (slide B) (see Figure 2.2).

☐4. With the *scanning objective* in position on the microscope, place slide A on the microscope stage so that the leaf is pointing away from you. Now look through the microscope and focus on the tip.

Q1. Is the leaf image seen through the microscope *right side up* (still pointing away from you) or is it *upside down* (now pointing toward you)?

_____

☐5. Remove slide A and replace it with slide B, positioned on the stage such that the leaf tip is pointing to your right. Now look through the microscope.

Q2. Is the leaf image seen through the microscope also pointing to the right, or is it *backwards* (pointing to the left)? _____

Q3. Thus, combining answers 1 and 2, you can say that the optical systems in a compound microscope produce an image that is _____

_____.

☐6. It takes some practice to move the slide correctly so that the image moves in the desired direction.

Q4. To move the image to the right, in which direction do you move the slide? _____

Q5. To move the image away from you, in which direction do you move the slide? _____

NOTE: **Dissecting microscopes** (used in other laboratory investigations to view large objects) are designed so that the image moves in the *same* direction that you move the object.

☐7. Using the proper procedure, switch from scanning to low power and focus the image; then switch to high power and focus again.

Q6. As you moved to higher and higher power objectives, did the brightness of the field of view get brighter, remain the same, or get dimmer?

_____

Does this make sense, considering that your field of view has been filled by a smaller and smaller area of the illuminated slide as you moved up in magnification?

direction of viewing the whole leaf

cover slip

cross section of leaf

the depth of the specimen that is in focus when viewed from above

microscope slide

Figure 2.3 Depth of field under high power

2

Q7.   What is the working distance, in millimeters, when the *high-power* objective is in focus?

_____

☐8. On high power, focus up and down through the *Elodea* leaf.

Q8.   How many cell layers thick is your *Elodea* leaf? (To determine this, focus through the specimen with the fine focus and observe whether there is no change in the image, or whether one "brick-pavement pattern" of cells is suddenly replaced by a different layer of brick-pavement pattern as you keep focusing.) _____

Q9.   **Depth of field** refers to the depth of the specimen that is in focus at a particular focus setting. Examine the depth of field for each objective. As you go up in objective power, is the depth of field decreased, the same, or increased?

_____

Q10. In Figure 2.3, indicate the depth of field under the *low-power* objective.

Q11. Experienced microscopists constantly move the fine focus up and down while viewing a specimen under a microscope. You should get into that habit, too. What is the benefit of doing this?

☐9. To gain an appreciation for the size of a field of view under different objectives, position one of your slides under scanning power such that a corner of the cover slip is in focus "at 9 o'clock" at the edge of the field of view. Then move along the edge of the cover slip, counting *the number of end-to-end fields of view that span the width of the cover slip* under scanning power. Repeat this procedure for the low-power and then the high-power objectives. In Table 2.1, record for each objective the number of end-to-end fields of view that span the width of the cover slip.

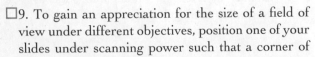

 PART II:
PLANT CELLS ARE DYNAMIC!

OBJECTIVES

1. Become familiar with the appearance of a cell at different focal levels under high power.
2. Observe the movement of chloroplasts by cytoplasmic streaming.
3. Learn that chloroplasts can be positioned by the cell to optimize photosynthesis.
4. Observe the dynamic movements of the vacuolar membrane in time-lapse videos.

BACKGROUND

A cell is a dynamic structure in which internal structures and materials are moved around (see Figure 2.4). The central region of a plant cell is occupied by a large membrane-bound organelle called the **vacuole**, which stores and recycles various materials of the cell. Under the light microscope, vacuoles look as if they are empty, hence the name (Latin: *vacuus* = empty). The **chloroplasts** (the green organelles within which photosynthesis occurs; Greek: *chloros* = green, greenish-yellow) are located in the cytoplasm between the vacuole and the **cell membrane**, which is just within the **cell wall**. With patient observation of a fresh *Elodea* leaf under a microscope, you can see the little pill-shaped, green chloroplasts being slowly moved around in the cytoplasm. The molecular mechanism responsible for **cytoplasmic streaming** is still being actively researched.

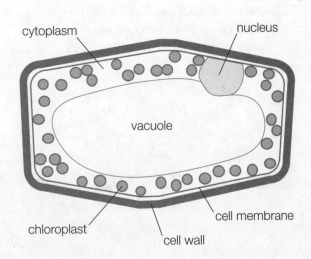

Figure 2.4 *Elodea* cell and some of its components

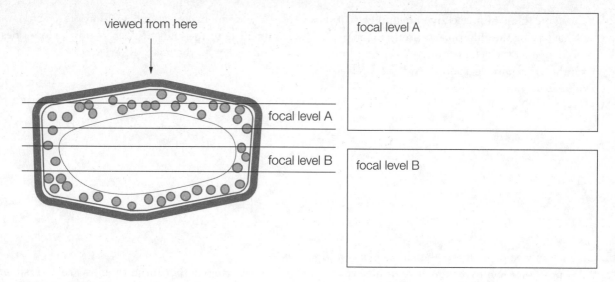

viewed from here

focal level A

focal level B

focal level A

focal level B

Figure 2.5 Focal levels at high power

MATERIALS

- microscope materials of Part I
- *Elodea*—living (not under strong light just before lab)
- Internet connection

INVESTIGATIONS

☐1. Pick a normal-sized, healthy *Elodea* leaf from the shoot tip; examine the cells of the terminal half of the leaf under scanning, then low, and finally high power. On high power, use fine focus to focus up and down through a cell. In Figure 2.5, draw the chloroplast arrangement seen at two different focal levels (A and B) when that "slice" of the cell is viewed from above.

☐2. Observe for several minutes the cells at the edge of a healthy *Elodea* leaf under high power, paying close attention to the positions of the chloroplasts. If the slide starts to dry out, add a drop of water to the edge of the cover slip. Note that chloroplast movement is *very slow*. If no movement is apparent, look again after a few minutes, continually keeping the microscope light on the leaf.

Q12. In what direction do the chloroplasts within one cell move? _____

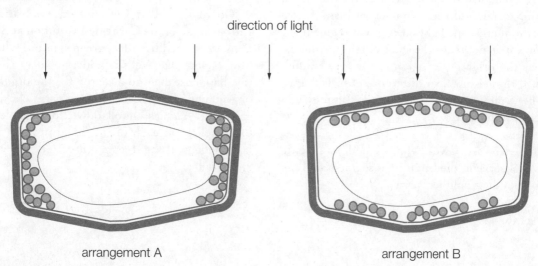

direction of light

arrangement A   arrangement B

Figure 2.6 Chloroplast arrangements

Q13. Can you think of a selective advantage to the cell of having the chloroplasts pretty much plastered against the cell membrane? (Hint: Photosynthesis requires carbon dioxide and releases oxygen.)

Q15. Under high power, what chloroplast arrangement is in the cells of each region? How does this relate to the color difference?

☐3. Cellular mechanisms move the chloroplasts and can also position them to maximize their absorption of light when in dim light; furthermore, *chloroplast avoidance movements* occur to minimize the exposure of chloroplasts to intense light. Intense light is damaging to the photosynthetic process in chloroplasts. Would you expect chloroplast arrangement A or B (Figure 2.6) to occur in dim light? _____ Would you expect chloroplast arrangement A or B to occur in intense light? _____ Explain your reasoning:

☐5. Determine whether these changes are reversible. Remove the slide from the microscope light and reexamine every 5 minutes.

Q16. On your slide, did the cells of the brightly lit region revert to match the cells of the surrounding area?_____
Was the effect of the bright light irreversible at any of the dimmer settings? Why might that be?

☐4. Pick a normal-sized healthy *Elodea* leaf from a shoot tip that is not in bright light; mount this leaf on a slide. Note the chloroplast distribution in the cells of the terminal half of the leaf. Now investigate the effect of bright microscope light on a segment of the terminal half of the leaf (different students in the group can use different dimmer settings). Leave the bright microscope light on that leaf segment for 10 minutes; do not move the slide. After 10 minutes, compare the brightly lit region to the surrounding region. If there is no clear difference observed, give it another 5 minutes and reexamine; do this until you see a clear difference.

Q14. Under lowest power, what color difference do you observe in the brightly lit segment and the surrounding unlit leaf area?

☐6. Go to the Plant Cell Imaging website of the Ehrhardt lab of the Carnegie Institution at Stanford University, at http://deepgreen.stanford.edu/index. html. (Note that website addresses may change; if this happens, then do a search for the organization or web page title.) On the home page, click on Organelle Dynamics; scroll down to Vacuole Membrane; click on each of the images to get to timelapse videos of the dynamic nature of the vacuolar membranes.

PART III:
DIVERSITY IN PLANT CELLS

1. Observe the diversity of chloroplast shapes.
2. Learn to distinguish between and identify raphides, druses, and cystoliths.
3. Observe and identify chromoplasts, oil droplets, and a diversity of starch grains and crystals.
4. After viewing an array of plant specimens, be able to use your knowledge to evaluate and identify an "unknown" selected from that array.

BACKGROUND

Forensic botanists who are familiar with the diversity of plant cellular structure can match or even identify small plant fragments. Although cell division produces cells that are at first quite similar, the process of **differentiation** produces a diversity of cellular forms. Cells can be small to very large, roundish to elongated to branched, thin-walled to thick-walled to stony, and with or without variously shaped chloroplasts, crystalline structures, starch grains, and so forth. Certain cellular forms are advantageous for particular functions that are needed in certain locations within the plant. For example, elongated, thick-walled cells in vascular tissue are optimal for conducting water up a plant; stony cells of a peach pit effectively protect the seed enclosed in the pit; thin-walled cells full of starch grains compactly store a large amount of energy in potatoes. Any genetic mutation that results in a cell structure that makes the cell more effective in a function will contribute to the plant's survival to reproduction, and thus will be passed on to the next generation.

III.A. CHLOROPLAST DIVERSITY

The chloroplast shape seen in *Elodea* is very common in plant leaves. Are chloroplasts shaped like little green pills in all photosynthetic cells? In this part of the lab, you will examine the shape of the green chloroplasts in several species of green algae.

MATERIALS

- microscope materials of Part I
- pipets with bulbs
- *Micrasterias*, living culture (Carolina HT-15-2345, Ward's 86 W 0270)
- *Mougeotia*, living culture (Carolina HT-15-2360, Ward's 86 W 0280)
- *Spirogyra*, living culture (Carolina HT-15-2525, Ward's 86 W 0650)
- *Zygnema*, living culture (Carolina HT-15-2695, Ward's 86 W 0900)

INVESTIGATIONS

☐1. You may be instructed to work in groups for this part. Make a wet mount of each algal culture and examine each under the microscope. Determine the shape of the chloroplast(s) in each.

Q17. Draw a line to connect the name with its chloroplast shape below:

Elodea	• Flat, long, corkscrew-shaped ribbon; one or more in a cell
Micrasterias	• Flat, straight; as wide as the cell; running the length of the cell
Mougeotia	• Girdle-like, wide band around the middle one-third of the cell
Spirogyra	• Large, lobed plate filling the contours of each half of the cell
Zygnema	• Little green pills; several per cell
	• No chloroplasts
	• Single, flat, U-shaped; as wide as the cell
	• Two star-shaped chloroplasts per cell

III.B. CELL DIVERSITY: CELL WALLS, CELL SHAPE, CELLULAR INCLUSIONS

BACKGROUND

Cell wall thickness and cell shape are additional clues that can be useful to a forensic botanist. Depending on the cell type, the cell wall may be thin to thick to downright stony. Cell shape can be roughly spherical to very elongated to star-shaped. These features are easily observed under the microscope.

Photosynthetic plant cells contain chloroplasts, but other internal structures may also be visible in certain cells of certain plant species. The following internal cellular structures can help a forensic botanist to identify or match plant fragments:

- **Chromoplasts.** Pill-shaped plastid organelles containing yellow, orange, or red carotenoid pigments; common in flower petals; also in some fruits, roots, and aging leaves (Greek: *chroma* = color)
- **Starch grains.** Composed of chains of glucose sugars; common compact structure for energy storage in plants; various shapes depending on the species
- **Crystals.** Vary in composition and shape: calcium oxalate can form bundles of needle-shaped **raphides** or compact crystalline clumps called **druses**, while calcium carbonate can form very large **cystoliths** located in oversized epidermal cells of certain species. (Greek: *kystis* = bladder, *lithos* = stone)
- **Oil droplets.** Compact energy storage

Permanent mounts of plant structures have been available from suppliers for several decades. Steps in producing these slides include staining with various chemicals to increase contrast and make structures more distinctive (nuclei and heavy cell walls are stained red, regular cell walls green, starch grains purple, for example). The stained specimen is sliced by machine into very thin sections, which are then mounted in a resinous balsam material. A machine-produced thin section is so thin that it shows you a *slice* through a cell.

Unstained hand-sliced fresh plant material, however, retains the natural colors of the plant parts, shows quite a lot of detail, and is better for viewing large structures. For example, the large, branched, star-shaped cells in water-lily leaves can be viewed in their entirety in handmade sections, whereas only a slice of these cells is present in the typical, thin, machine sections on prepared slides. In this lab, you will hand-section some of the specimens as well as look at some prepared slides.

The first specimens that you will be examining will be explained to you; subsequent specimens you will have to evaluate on your own, based on what you have learned. Finally, you will apply your newly acquired knowledge to identifying several "unknowns"—unlabeled slides.

MATERIALS

- microscope materials of Part I
- paper towel
- *Medicago*, alfalfa stem, cross section (Carolina HT-30-2780, Ward's 91 W 8010)
- *Phormium*, New Zealand flax, leaf cross section (Triarch 16-9, www.triarchmicroslides.com)
- *Yucca* leaf, cross section (Carolina HT-30-4114, Ward's 91 W 7421)
- new, single-edge razor blades
- fresh water-lily leaf
- *Nymphaea*, water-lily leaf, cross section (Carolina HT-30-3892, Ward's 91 W 8061)
- pear and apple fruits
- living *Zebrina*, *Begonia*, *Euphorbia*, *Impatiens*, and *Ficus elastica* (rubber tree leaves)
- banana, parsnip, white potato, ginger "root," red sweet pepper, and red cabbage
- dropper bottle of IKI solution (1% iodine, 2% potassium iodide—laboratory grade)

INVESTIGATIONS

NOTES:
1. If the paper-thin slices that you are trying to cut are too thick to transmit light, try cutting a thin *wedge-shaped* slice and looking at the tapered edge of that wedge.
2. You may be instructed to work in groups for this part, examining each other's slides.

Diversity of Cell Shape and Cell Walls

☐1. Alfalfa. Examine the prepared slide of the cross section of an alfalfa stem. Note the diversity of cell diameters and cell wall thicknesses as you scan from the outside to the central pith region. The safranin O stain has stained the heavy cell walls of the water-conducting cells red. Can you find these water-conducting "xylem" cells?

☐2. Paper. Examine the edge of a small piece of paper towel that you have mounted in a drop of water on a slide. Draw a fiber in Worksheet 1.

14

paper fibers	yucca	water-lily leaf: branched cells	pear: cluster of stone cells	apple

2

☐3. Fibers. Examine the prepared slide of the leaf of *Phormium*, a plant from which fibers are extracted. Find the bands of thick-walled, red-stained fiber cells (seen in cross section) that extend across the leaf at regular intervals in this species. Other plants have other amounts and arrangements of fiber cells in their leaves.

☐4. Fibers. Examine the prepared slide of a leaf cross section of *Yucca*—another plant that is used as a fiber source. Locate the bunches of fiber cells—very thick-walled and stained red. In Worksheet 1, draw a portion of the Yucca leaf, labeling the fiber bundles.

☐5. Branched cells. Lay the blade of a fresh water-lily leaf flat on the table top. With a single-edge razor blade, cut a small square out of the leaf blade; then carefully cut several paper-thin sections from one edge of the square; mount the sections in a drop of water on a slide. Locate the large transparent star-shaped, branched cells in the leaf. Draw the branched cells in Worksheet 1.

Look at the *prepared, stained* slide of the water-lily leaf and notice that only a slice of each thick-walled, red-stained branched cell is contained within the thin section; being stained red, they are easier to pick out, but you get only a slice of them.

☐6. Stony cells. Cut and mount a paper-thin section of a pear and an apple, cutting down through the skin into the underlying flesh. Pears have a gritty texture due to scattered clusters of stony cells surrounded by radiating large cells, giving a daisy-like appearance. Draw what you see, on Worksheet 1.

Compare this pear section to a paper-thin section of an apple. Could pear and apple fragments be distinguished by a forensic botanist?

Diversity of Cellular Inclusions

☐7. Starch grains. Cut a wedge-shaped, paper-thin slice of white potato; as always, mount it in a drop of water and add the cover slip. Examine the tapered edge to locate transparent, egg-shaped starch grains released from ruptured cells or contained within intact cells. Add a drop or two of iodine to one edge of the cover slip; observe as the iodine seeps under the cover slip and stains the starch grains from purple to black. Draw the potato starch grains in Worksheet 2.

☐8. Diverse starch grains. For each of the specimens listed, prepare a slide, stain for starch, and draw the shape and relative size in Worksheet 2 (if the grain has fine "growth rings" within it, draw those also). Note that one of these plants has little dog-biscuit-shaped starch grains!

- Paper-thin slice or wedge of parsnip root
- Paper-thin slice or wedge of ginger "root," which is actually an underground storage stem
- A bit of mashed flesh of banana, spread thinly in water
- The white milky latex from *Euphorbia*: Place one drop of latex next to one drop of iodine solution on a dry slide, and cover both with the same cover slip; *do not add any water to the slide*; the iodine will infiltrate the latex and stain the starch grains.

WORKSHEET 2. Diversity of Starch Grains

Draw the shape and relative size of the starch grains of each species:

white potato	parsnip	ginger	banana	*Euphorbia*

WORKSHEET 3. Crystalline Structures Observed

Draw the crystalline structures seen in each specimen:

Species and Plant Part	Raphides?	Druses?	Cystoliths?

☐9. Oil droplets. Examine an unstained, paper-thin wedge of ginger, looking for greenish-gold oil droplets in scattered cells along the tapered edge of the wedge.

☐10. Crystals. Raphides, druses, or cystoliths are present in these listed plants; examine each, determine the crystalline structures present, and draw them in Worksheet 3.
- Paper-thin slice of fresh *Begonia* stem — carefully examine cells of the central region; once you've spotted one crystal, you will be able to spot the others more easily.
- The mashed end of a *Zebrina* stem
- Paper-thin slice of a *Ficus elastica* leaf blade piece
- Paper-thin slice or mashed end of *Impatiens* stem

☐11. Chromoplasts. Some plant structures are red due to red pigments in chromoplasts, and some are red due to red pigments in the large vacuoles. Prepare and examine sections of red sweet pepper and red cabbage.

Q18. Chromoplasts are present in _____.

☐12. Now be a forensic botanist! Your instructor will have three microscopes with unlabeled slides of specimens that you have examined during this lab period. Examine the unknowns and based on your experience today, identify the unknowns. Record your identifications in Worksheet 4.

FURTHER INVESTIGATIONS

Students can make fresh mounts of paper-thin slices taken from various portions of other plants, fruits, and vegetables, evaluating them for cellular inclusions. Based on this array of plants examined, does a particular shape of crystal or starch grain appear to occur more commonly than others do? In a plant that has raphides, do they occur throughout the plant, or are they localized in certain parts only? Which of these plants are quite distinctive under the microscope and distinguishable from the other plants examined in this lab? Do species that are closely related have the same types of cellular inclusions?

WORKSHEET 4. Identifying Unknowns

Unknown	Identified as Plant	Based on What Features
1		
2		
3		

16

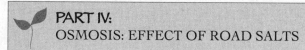

PART IV:
OSMOSIS: EFFECT OF ROAD SALTS

OBJECTIVES

1. Understand the direction in which water will move across a membrane by osmosis.
2. Understand the relative position of and lack of connection between the cell membrane and cell wall.
3. Understand how salt affects cells and thus damages plants.

BACKGROUND

Salt runoff or salt spray from roadways and oceans is damaging to most plants. Only certain plants have special physiological adaptations that enable them to grow in salt marshes or in areas exposed to ocean spray. Salinization of soil is a problem in irrigated fields—salts from the water and fertilizer gradually build up in the soil over hundreds of years until the salt concentration is so high that crops can no longer grow there (this led to the decline of several ancient civilizations). Researchers are now actively investigating salt tolerance and working to produce more salt-tolerant crop varieties.

In this part of the lab, you will investigate the effect of various salt concentrations and various kinds of road salts on the leaf cells of the freshwater plant *Elodea*.

Lab 6 (Seed Germination) includes an investigation of the effects of salt on seed germination.

Some terminology that is relevant: A **solution** consists of the liquid **solvent** in which the **solute** material is dissolved; the greater the solute concentration in a solution, the less the water concentration is in that solution. **Diffusion** is the passive movement of something from a region where it is in higher concentration to a region where it is in lower concentration. **Osmosis** is the diffusion of water across a membrane, moving to the side that has a lower water concentration. Water moves rather freely across cell membranes. A cell that has lost water is referred to as **plasmolyzed**; it has undergone plasmolysis.

When the solutions on each side of a membrane have the *same* concentration of dissolved material (solute), the concentration of water in the solutions is also the same. In this situation, an equal amount of water moves out of the cell as into it, resulting in *no net movement* of water across the membrane.

If solution A on the outside of the plasma membrane has a *lesser solute* concentration and thus *greater water* concentration than is present inside the cell, there will be a *net movement of water from the region of greater water concentration* outside the cell to the region of lesser water concentration inside the cell. This net movement continues until equilibrium is attained; equilibrium occurs when (1) the two solutions have reached the same solute concentration, or (2) in plant cells, when the rigid cell wall exerts enough turgor pressure to pre-

WORKSHEET 5. Appearance of *Elodea* Cells in Various Salt Solutions

	0% solution	1% solution	2% solution	3% solution	4% solution	5% solution
Cells in NaCl solutions						
Fully reversible?						
Cells in road salt _____ solutions						
Fully reversible?						
Cells in road salt _____ solutions						
Fully reversible?						

vent any more net water flow into the cell although the outside and inside solutions are not yet at equal solute concentrations (this is similar to the wall of an inflated bicycle tire resisting your efforts to pump more air into it).

Understanding these processes is what is important here. Yes, there is terminology if you desire it: Two solutions on either side of a membrane that have the same solute concentration are called *isotonic*. A *hypotonic* solution has a lower solute concentration than does the solution on the other side of a membrane. A *hypertonic* solution has a greater solute concentration than does the solution on the other side of a membrane.

Q19. If a cell is placed into a solution containing *more* dissolved material (a greater solute concentration) than the cell's cytoplasm has, would you expect any net movement of water? _____ In what direction? _____

Q20. (Plant cells have a cell wall surrounding but not attached to the cell membrane.) What do you expect will happen to a plant cell in a solution having a greater solute concentration than the cytoplasm? _____

MATERIALS

- microscope materials of Part I
- paper towels
- dropper bottles of 0%, 1%, 2%, 3%, 4%, 5% NaCl solutions (5% = 5 g NaCl in 100-ml solution)
- various kinds of road salts, brought in by students and staff (*NOTE*: Some kinds of salt crystals are very hygroscopic, absorbing much moisture from the air; within a few days, a paper bag of such crystals will be sitting in a pool of liquid. Therefore, keep your salts in *sealed plastic bags or jars* until use.)
- balances, weighing paper, and spatulas
- several graduated cylinders, 50 or 100 ml
- flasks in which to mix the road salt solutions
- pipets with bulbs
- *Elodea*—living
- tape

INVESTIGATIONS

☐1. You may be instructed to work in groups. Get 6 equal-aged *Elodea* leaves and lay each leaf in a few drops of 0%, 1%, 2%, 3%, 4%, or 5% NaCl on microscope slides. Use tape to label the slides.

☐2. After several minutes, add cover slips and examine each leaf under the microscope.

☐3. Draw the cells and their contents as they appear in each salt solution, on Worksheet 5.

☐4. Remove each leaf, blot it, and lay each in a few drops of distilled water on microscope slides; add cover slips and observe under the microscope.

☐5. Record your observations in the space provided.

☐6. Now repeat this investigation with various road salts that are available, mixing the salt solutions yourself. (The class can work together to mix a set of solutions for everyone to use.) Note that "5% NaCl" means 5 g NaCl in 100 ml, or 2.5 g in 50 ml, or 0.5 g in 10 ml of solution. As before, record your results on Worksheet 5.

Answer these questions:

Q21. Of the salts tested, is there one that you found to be clearly less damaging to *Elodea* than the others are? _____ Did the class as a whole find this to be so? _____

Q22. If a normal *Elodea* plant cell is put directly into distilled water (pure water), explain what will happen; also explain what the equilibrium situation will consist of.

Q23. If an animal cell (lacking a cell wall) were placed into distilled water, what would happen to it?

LITERATURE CITED

Cutler, S., and D. Ehrhardt. 2006. "Plant Cell Imaging," http://deepgreen.stanford.edu/index.html (accessed February 19, 2006). Stanford, CA: Carnegie Institution, Stanford University.

LAB

PHOTOSYNTHESIS: Pigments and Starch

I. Pigments in plant leaves
II. Starch in plants

1. Understand photosynthesis: be able to write the overall reaction giving its inputs and outputs.
2. Understand which pigments are involved in photosynthesis, what portion of the light spectrum each absorbs, and how each contributes to photosynthesis.
3. Understand the technique of paper chromatography and how it can be affected by various solvents; understand how scientists develop procedural details for a technique (Part I.A).
4. Understand the array of pigments and their function in a green leaf, in a red leaf of a red ornamental, and in a colorful autumn or other dying leaf (Parts I.A, I.B, and I.C).
5. Understand the relationship between sugar (glucose) and starch; observe starch grains (Part II.A).
6. Understand the starch content in healthy leaves: a green leaf, a red leaf from a red ornamental, and a variegated leaf (Part II.B).
7. Understand the changes in starch content when a leaf is dying (Part II.B).

BACKGROUND

Photosynthesis Requires Pigments and Produces Sugars and Starch

Green plants have green leaves; the leaves are green because of the green pigment called **chlorophyll** (Greek: *chloros* = green, greenish-yellow; *phyllon* = leaf), which is involved in photosynthesis (Greek: *photo* = light). Let's look at what is required for efficient photosynthesis.

A leaf has evolved, chemically and structurally, to optimize photosynthesis. The overall function of the biochemical process of **photosynthesis** is to absorb light energy and convert it into chemical bond energy that is then usable by the plant. This chemical bond energy is within the glucose sugar, which is synthesized by the photosynthetic process. Thus, it is sometimes said that a plant gets its "food" (glucose) from sunlight. The "inputs" required by photosynthesis are light, carbon dioxide, and water. The "outputs" produced are glucose, oxygen, and water. Your textbook provides greater detail of the biochemical process involved in photosynthesis.

$$6CO_2 + 12H_2O + LIGHT \xrightarrow{\text{chlorophyll}} C_6H_{12}O_6 + 6O_2 + 6H_2O$$

solar energy

glucose
chemical bond energy

Let's focus on **light** and its capture by a cell. The visible light spectrum ranges from red (the longest wavelength) through orange, yellow, green, blue, indigo, and finally violet (the shortest wavelength). The green pigment that absorbs light and is directly involved in photosynthesis is called **chlorophyll *a***. Chlorophyll *a* absorbs violet/blue and red light very readily but not much of the lighter blue and the green and yellow light (see Figure 3.1). The lighter blue, green, and yellow light that is not well absorbed by chlorophyll *a* is reflected back to our eyes, and that is why chlorophyll *a* looks *bluish green*.

In addition to producing chlorophyll *a*, leaves have evolved to produce several other pigments collectively termed **accessory pigments**; accessory pigments also absorb solar energy, but they then pass this absorbed

NOTE: Your instructor may have the class do only certain parts of this lab or may have different student groups do different parts and then report their results to the class.

19

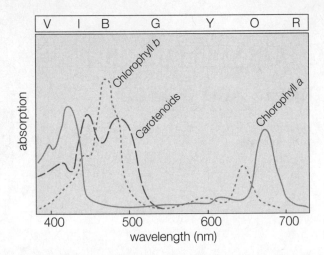

Figure 3.1 Absorption spectrum of chlorophyll *a*, chlorophyll *b*, and carotenoids. The letters VIBGYOR refer to the colors violet, indigo, blue, green, yellow, orange, red.

energy over to chlorophyll *a* for photosynthesis. **Chlorophyll *b*** is structurally only slightly different from chlorophyll *a*, but its absorption spectrum is somewhat different. Chlorophyll *b* absorbs more of the blue range along with the red (Figure 3.1); thus, chlorophyll *b* reflects mostly green and yellow light and looks to our eyes yellow-green. There are also a variety of accessory pigments collectively called **carotenoids** (Latin: *carota* = carrot); these carotenoid pigments in leaves absorb a great deal in the blue and green range, and thus appear to our eyes as various shades of *yellow* or *yellow-orange*.

Why would it be a selective advantage to have accessory pigments? These accessory pigments can absorb wavelengths of light that chlorophyll *a* cannot absorb effectively, and thus the plant is able to use more of the sun's energy than it could with chlorophyll *a* alone. Note that in addition to the pigments that are involved directly or indirectly in photosynthesis, plants can also produce other kinds of pigments. For example, pigments in flower petals function in attracting particular pollinators, while pigments in ripening fruit function in attracting hungry animals that then serve in seed dispersal.

Leaf Structure Optimizes Photosynthesis

Leaf **structure** has evolved to optimize photosynthesis in its cells. Leaves are *thin* so that the light can penetrate and reach all the photosynthetic cells within the leaf. Leaves of land plants are covered with a waxy **cuticle** layer to prevent water loss, but the cuticle also has the negative effect of preventing gas exchange. Therefore, the outer cell layer (epidermis) also has pores called **stomata** (singular: stoma; Greek: *stoma*

= mouth) through which gas exchange can occur by "diffusion." **Diffusion** is the movement of molecules from a region of higher concentration to a region of lower concentration. (The fragrance of a rose diffuses through the air, being strongest nearest the rose.) Carbon dioxide that is needed in photosynthesis travels by diffusion through stomata into the leaf, and excess oxygen that is produced by photosynthesis travels by diffusion out of the leaf. Another structural feature is the **air spaces** within the leaf that permit diffusion of gases throughout the interior of the leaf to reach all the cells. Leaves also have veins of **vascular tissue** that conduct water and minerals to all the leaf cells and conduct excess energy-rich sugar from the photosynthetic cells to other plant parts for use or storage. If glucose molecules are to be *stored* for later use as an energy source, they are chemically bonded to each other and the resulting structure is called **starch**. We get energy from eating starchy french fries by breaking down the large starch grains within the potato cells.

Today you will determine the kind of chromatography solvent that gives the best separation of spinach-leaf pigments; you will then use that solvent to investigate the pigment content of nongreen leaves and of dying leaves. In addition, you will stain for starch in various healthy leaves and in dying leaves. If the instructor assigns different parts to different student groups, your results will be shared by oral presentation.

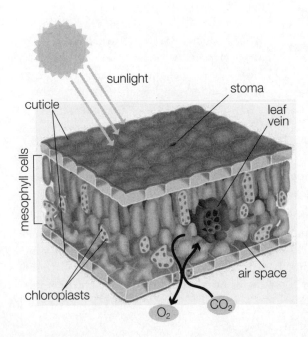

Figure 3.2 Cross section of a leaf. CO_2 and O_2 are passing through a stoma

PART I:
PIGMENTS IN PLANT LEAVES

I.A. IN SPINACH, ARE THERE OTHER PIGMENTS BESIDES GREEN ONES? ALSO, WHICH CHROMATOGRAPHY SOLVENT GIVES THE BEST PIGMENT SEPARATION?

BACKGROUND: WHAT IS CHROMATOGRAPHY?

The method of **paper chromatography** can be used to separate the components of a mixture of molecules—a mixture of pigments or a mixture of amino acids, for example. The mixture is spotted onto a solid support (paper); the paper's end is put into a solvent; as the solvent travels up the paper, the different types of molecules of the mixture are carried along at a rate that is determined by (1) the molecule's affinity for the solid (paper), and (2) the molecule's solubility in the solvent. At the end of a chromatography run, the mixture will have separated into a series of spots *if* the different components of the mixture differ in their affinity and solubility properties. Note that if two different pigments happen to have the *same* affinity and solubility properties, then they travel at the *same* rate and will be in the *same* spot on the completed strip, now called a **chromatogram**. Chromatography of leaf pigments results in an array of pigment spots that are colored and visible directly. (In contrast, chromatography of amino acids results in spots that are invisible on the chromatogram and must first be dried and then sprayed with a reagent that reacts with the amino acids to produce color.)

There are various leaf pigment chromatography procedures, specifying various ratios of acetone to petroleum ether for the solvent. Does the solvent ratio make much difference? Is there an optimal solvent ratio?

In this experiment, you will see the work involved in perfecting a lab technique; you will also observe the various pigments present in green leaves. To do this, you will prepare seven paper strips for chromatography by applying spinach-leaf pigments onto the strip, and then use seven different solvents in the tubes for the running of the chromatography. The purpose of this experiment is twofold: (1) to determine the effect, if any, that the different solvents have on the outcome, and (2) to determine the best solvent to use for optimal pigment separation.

MATERIALS

Per group:
- 7 large-diameter test tubes (25 × 200 mm) in a test tube rack
- chromatography paper, Whatman 3MM Chr (Whatman No. 3030-614, 20 mm × 100 m roll)
- scissors
- solid neoprene rubber stoppers to fit tubes
- fresh spinach leaves
- white tape and marker
- ruler

In a hood or well ventilated area:
- distilled water
- 0% acetone, 100% petroleum ether solution
- 4% acetone, 96% petroleum ether solution
- 8% acetone, 92% petroleum ether solution
- 12% acetone, 88% petroleum ether solution
- 16% acetone, 84% petroleum ether solution
- 20% acetone, 80% petroleum ether solution
- 5-ml pipet and labeled pipet pump for each solution
- empty glass bottle labeled "organic wastes"

INVESTIGATION (work in groups of two today)

A NOTE ON SAFETY: Both acetone and petroleum ether are irritating to the eyes, skin, and respiratory tract. In addition, both substances are extremely flammable. Use them in a well-ventilated area such as a fume hood, and keep them away from sources of ignition. Wear safety glasses. See the supplier's "Material Safety Data Sheet" for more details. (You may recognize the smell of acetone since it is the main ingredient of regular fingernail polish remover.)

☐1. Prepare your seven large-diameter test tubes as indicated below if they have not already been filled for you by the staff; use *separate pipets and pipet pumps* to put 5 ml of the appropriate solution into each tube; lightly stopper each tube; be careful to correctly label each tube with the solvent ratio it contains. **(Remember: Do not breathe the fumes, and keep the solvents away from ignition sources!)**

    Tube 1: distilled water
    Tube 2: 0% acetone, 100% petroleum ether
    Tube 3: 4% acetone, 96% petroleum ether
    Tube 4: 8% acetone, 92% petroleum ether
    Tube 5: 12% acetone, 88% petroleum ether
    Tube 6: 16% acetone, 84% petroleum ether
    Tube 7: 20% acetone, 80% petroleum ether

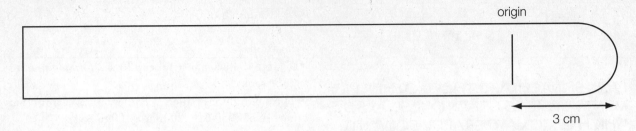

origin

3 cm

Figure 3.3 Chromatography strip

☐2. Cut seven strips of chromatography paper to a length that fits inside your stoppered test tubes.

   a. Round the bottom of each strip, and lightly draw a pencil line to mark the "origin" across each strip, about **3 cm** from the rounded end (see Figure 3.3).

   b. Label the top of each strip with one of the solutions listed in step 1.

☐3. Lay a spinach leaf onto the end of the paper strip and transfer a line of leaf pigments onto the origin by using a rounded hard edge (coin, scissor, or vial edge) to press down and crush the leaf along the origin line. You want a line of extract that is not touching the edges of the strip.

☐4. Repeat until you have a dark line of pigment on the paper; the line can be 2–3 mm wide. The pigment line on the strip need not dry.

☐5. Insert each strip into its corresponding tube. The bottom tip of each strip must be immersed, but be sure that **the line of pigments is never immersed in the solution**.

☐6. Lightly stopper each tube; set in a rack in a well-ventilated area.

☐7. Double-check to make sure that the label on the strip and its tube match.

☐8. Let the tubes stand *undisturbed and vertically* in the rack. (While you wait, gather the leaves and prepare the paper strips needed for the next experiments.)

☐9. Monitor the progress, and remove each strip **when the leading edge of the solvent is about 1 cm from the top of the strip** (roughly 30 minutes); lay the strip, now called a chromatogram, on a paper towel to dry in a well-ventilated area.

☐10. Pigment spots fade somewhat when dry; before each strip is fully dry, use a pencil to outline each pigment spot precisely. Your pigment spots will be in the shape of lines or bands. When looking for pigment spots, don't forget to check at the leading edge of the solvent.

☐11. **Do not discard your solvents down the sink!** Discard the solvents into the "organic wastes" container indicated to you, for proper disposal by your institution.

☐12. Tape the strips by their tops and bottoms (in sequence, from tube 1 to 7) onto your group's hand-in sheet, which is at the end of this lab.

☐13. *Label each pigment spot* as to whether it is **chlorophyll a** (blue-green), **chlorophyll b** (yellow-green), or **carotenoids** (yellow or yellow-orange).

Q1.   Did the type of solvent make any difference in the separation of the leaf pigments? _____ What is the optimal solvent for separating spinach leaf pigments farthest from each other?

Q2.   At least how many pigments are in spinach leaves? _____

Q3.   Could there be more than one pigment per pigment spot? Why? (Reread the background paragraph on "What Is Chromatography?")

I.B. WHAT PIGMENTS ARE IN RED CULTIVAR LEAVES?

Some ornamental plants such as Crimson King Norway maple, purple beech, purple-leaf plum, and purple-leaf sand cherry have been bred to have dark red or purple leaves all summer long. These cultivated varieties, or "cultivars," are planted to provide a contrast in the summer garden. Red varieties have also been bred for some houseplants such as *Coleus*.

The purpose of this experiment is to see whether red cultivar leaves contain pigments other than red ones. (Green plants perform photosynthesis; how can red plants live?)

MATERIALS

- red-leaved ornamental tree or shrub on campus
- winter alternative: a red-leaved variety of a tender-leaved houseplant such as *Coleus*
- the nonleaf materials from Part I.A.

INVESTIGATION

☐1. Obtain a leaf of a red-leaved cultivar.

☐2. Repeat the extraction and chromatography procedure of Part I.A. For running the chromatography, use the optimal solvent as determined by your results in Part I.A.

Q4. Which pigments did the green leaves and red cultivar leaves have in common?

Explain the function of the various pigments that you see in the red cultivar leaves.

I.C. WHAT PIGMENTS ARE IN AUTUMN OR SENESCENT LEAVES?

The purpose of this experiment is to see what happens to the leaf pigments when the leaf is **senescing** (declining and dying). (Latin: *senescere* = to grow old)

MATERIALS

- tree on campus with deep-green as well as colorful autumn leaves
- winter alternative: tender-leaved houseplant with green as well as dying yellow (formerly green) leaves
- the nonleaf materials from Part I.A.

INVESTIGATION

☐1. Gather a deep-green leaf and a nongreen senescent (declining, dying) leaf from an autumn tree or from a houseplant.

☐2. Repeat the extraction and chromatography procedure as in Part I.A., using the optimal solvent.

Q5. What changes do you see in the pigment composition of autumn or other senescent leaves as compared to the green leaves from the same plant? Why might this be happening?

Q6. Would you *expect* there to be a difference in the pigment content of nonsenescing *red* cultivar leaves and *red* autumn leaves? Explain why. (If you were able to perform pigment chromatography on both of these kinds of leaves, did you observe the results that you expected?)

DURING WAITING TIMES, DO THE FOLLOWING PART.

BACKGROUND: STARCH IS AN EFFICIENT WAY FOR PLANTS TO STORE GLUCOSE

The excess glucose sugars produced by photosynthesis can be stored in the leaf cells, or they can be transported via the vascular system to be stored elsewhere. When glucose sugars are to be stored in a plant cell, they are bonded together to form molecules called **starch**; the starch molecules in turn form starch grains in the cells. Some plants have structures specialized for starch storage; white potatoes are starchy underground storage stems on the white potato plant.

Deciduous plants (Latin: *decidere* = to fall off) drop all their leaves at the start of the inhospitable season (before winter in temperate regions; before the dry, hot summer in desert regions). This might seem like a terrible waste, after the plant has spent all that energy making the leaves. It is not, however, because during the process of leaf **senescence**, the plant breaks down whatever it can within the leaf and transports it into the twigs and stems in preparation for dropping the leaf. (It's sort of like stripping a car before dumping it.) Thus starch is broken down and removed before leaf drop; the chlorophylls are also broken down, revealing the underlying carotenoid accessory pigments along with other pigments in the colorful autumn leaves. Note that the leaves that fall to the ground are not a waste—they serve as an insulating layer for the roots and host a whole community of organisms. In addition, once the leaves are decomposed, they serve to fertilize the soil and thus nourish plants.

II.A. STARCH GRAINS IN POTATOES

The purpose of this section is to observe starch grains and learn the starch staining technique.

MATERIALS

- white potato
- single-edge razor blades
per lab table:
- dropper bottle of water
- dropper bottle of IKI solution (1% iodine, 2% potassium iodide—laboratory grade)
- microscope slides and cover slips
- compound microscope, 1 per student

- lens paper
- tissues or paper towels

INVESTIGATION

☐ 1. Using a razor blade, slice an extremely thin wedge of tissue from a white potato, place it in a drop of water on a microscope slide, cover it with a cover slip, and look at it with a compound microscope. Refer to Lab 2 (Microscopy and Plant Cells) for instructions on using a microscope.

☐ 2. Look for transparent, egg-shaped starch grains that have spilled out of ruptured cells along the thinnest edge of the wedge. (Other kinds of plants have other shapes of starch grains.)

☐ 3. Carefully add a drop of iodine solution (IKI solution) to one edge of the cover slip; iodine stains starch a deep purplish black. (**NOTE:** *Iodine also stains clothes! Wipe up spills immediately.*)

☐ 4. Look through the microscope and watch the clear, egg-shaped starch grains turn purple as the iodine solution reaches them.

II.B. STARCH IN A GREEN LEAF VS. RED CULTIVAR, VARIEGATED, AND AUTUMN OR OTHER SENESCENT LEAF

The purpose of this section is to observe the effect of excessive red pigment, variegation, or senescence on the presence and distribution of starch in leaves.

MATERIALS

- green leaf (*Coleus* variety, or other indoor or outdoor plant such as maple)
- red cultivar leaf (*Coleus* variety, or other indoor or outdoor plant)
- green-and-white variegated leaf (*Coleus* variety, or other indoor or outdoor plant)
- in autumn: tree on campus with deep-green as well as colorful autumn leaves
- alternative: tender-leafed houseplant with green as well as dying yellow (formerly green) leaves
- forceps
- two hot plates, placed far away from the acetone and petroleum ether chromatography work
- aluminum foil
- large Pyrex® beaker of distilled water, covered with aluminum foil; on one of the hot plates
- large Pyrex beaker of ethanol, covered with aluminum foil, *placed in a Pyrex* or *metal pan* of water on the other hot plate; extra foil available

- beaker of tap water
- IKI solution in a glass pan or large petri dish
- paper towels

INVESTIGATIONS

☐1. Get a deep-green leaf, a red cultivar leaf, a variegated leaf; also from a plant in autumn or a houseplant with some dying leaves, get a *deep green* leaf and a non-green senescent leaf. Maple leaves or others that are not very tough work well.

☐2. Draw the outline of each leaf in the area provided on the next page; mark and label any color patches on the leaf.

☐3. Boil the leaves in a foil-covered water bath on a hot plate for about 3 minutes to rupture the cells. The foil cover reduces evaporation.

☐4. With forceps, *carefully* transfer the leaf to a foil-covered *slowly boiling* ethanol bath *sitting in a pan of water* on a hot plate. Boil in ethanol to extract leaf pigments—about 6 minutes, until the leaf is whitish. (The aluminum foil cover reduces evaporation and also is handy in the unlikely event that the ethanol bath catches fire.)

☐5. With forceps, *carefully* dunk the leaf into tap water briefly and then transfer the leaf to a dish containing IKI iodine solution. Leaf areas with starch will turn purplish black. If you wish, you can remove excess stain by placing the leaf into the beaker of tap water.

☐6. Compare your staining patterns to those of other student groups to get a class consensus.

Q7. What is the *class consensus* for the starch staining pattern for each kind of leaf investigated?

green:

red cultivar:

variegated:

autumn or senescent:

Q8. Explain the presence and location of starch in your red cultivar leaf as compared to your green leaf.

Q9. What can you conclude about the deposition of starch grains relative to the site of glucose synthesis in variegated leaves?

Q10. How do the white non-photosynthetic regions of the variegated leaf remain alive (get energy)?

Q11. Compare the starch staining of your deep-green to your autumn or senescent leaves; what is revealed about the processes occurring in preparation for leaf drop? Discuss.

EXTENDED ASSIGNMENT

Select one of the parts of this lab, and design a future experiment or two that you could do to further clarify the subject. Indicate the question or hypothesis that you would be investigating and then describe the experimental design.

Leaf before Staining	Leaf after Staining, Showing Starch Areas
Solid-green leaf: draw outline	
Red cultivar leaf - is the red evenly distributed? ___	
Variegated leaf - precisely indicate and label the green and white regions.	
Autumn or senescent leaf - precisely indicate and label the various colored patches. Deep green leaf from the same plant: draw outline	

PHOTOSYNTHESIS LAB: Chromatography Results (Hand-in Sheet)

NAMES of group members:_____

Below, tape your seven labeled chromatograms *in order*; then carefully *outline each pigment spot* with pencil or pen, and *label each pigment spot* (**chl *a*, chl *b*, carotenoid**).

3

ACKNOWLEDGEMENT

The earliest mention of transferring leaf pigments directly onto paper by crushing the leaf on the paper, was found by Schimpf and Monson (1987) to be in Knaphus (1967).

LITERATURE CITED

Knaphus, G. 1967. *Botany laboratory manual, preliminary edition*. Dubuque, IA: Wm. C. Brown.

Schimpf, D. J., and P. H. Monson. 1987. How-to-do-it: A shortcut to leaf pigment chromatography. *The American Biology Teacher* 49(8):427–8.

DNA AND MUTATIONS

LAB

4

I. DNA replication
II. From gene to protein
III. DNA mutation and its effect on protein structure
IV. DNA mutation and its effect on protein function
V. DNA mutation and its effect on the organism

4

OBJECTIVES

1. Know the basic structure of DNA—how nitrogenous bases are arranged in a single strand of DNA and how two strands are held together in a double helix.
2. Know how DNA is replicated and how the replication process relies on base complementarity.
3. Given the base sequence of a DNA template strand, be able to transcribe it into the base sequence of its mRNA copy, and then use the genetic code table to translate this mRNA into the amino acid sequence that is encoded.
4. Given the base sequence of the normal and a mutated form of a gene, be able to determine the amino acid sequence of the normal and the mutant proteins.
5. Be able to compare a mutant protein to the normal protein and hypothesize the effect of the altered amino acid sequence on the mutant protein's three-dimensional structure and function.
6. Understand the possible effects that a mutation may have on a cell's or organism's normal function.

BACKGROUND

Deoxyribonucleic acid, abbreviated **DNA**, is the hereditary or genetic material of a cell. The DNA molecule is a "double helix" consisting of two strands. A DNA strand is a chain of **nucleotides** held together by strong covalent chemical bonds. Each nucleotide consists of a **phosphate** group bonded to a **deoxyribose** sugar, which is bonded to a **nitrogenous base**. The deoxyribose sugar is also bonded to the next nucleotide's phosphate group. The result is that a DNA strand has a phosphate-sugar-phosphate-sugar backbone, and a nitrogenous base is attached to and jutting out from each sugar of the backbone. The two strands are held together in a double helix due to hydrogen bonds between the nitrogenous bases. This gives the double helix the overall appearance of a ladder with the rungs being the base pairs (see Figure 4.1).

DNA has four kinds of nitrogenous bases: **adenine** (A) and **guanine** (G), each having a large two-ring structure, and **cytosine** (C) and **thymine** (T), each having a small one-ring structure. Because of the chemical structural details, adenine can hydrogen bond only to thymine, forming an **A-T base pair**, while guanine can hydrogen bond only to cytosine, forming a **G-C base pair**. Thus the two strands of a double helix are said to be **complementary**, due to the complementary base pairing.

The linear array of bases or **base sequence** along a strand of DNA is what is encoding genetic information, as we will see.

For simplicity, the three-dimensional DNA double-helix structure can be drawn using a line for each phosphate-sugar backbone and a letter abbreviation for each base:

The DNA double helix consists of two helical strands stably attached to each other by many weak hydrogen bonds. This is analogous to two pieces of Velcro® held together by many weak hooks.

In preparation for cell division, the DNA replicates itself. First the hydrogen bonds between the two strands break, and then a new complementary strand is synthesized along each original strand. The result is two double-helix molecules that are identical in base sequence; each double helix consists of one original strand and one complementary, newly synthesized strand.

Practice 1:

Replicate the DNA in Figure 4.2 by writing in a complementary strand alongside each original strand that serves as a template.

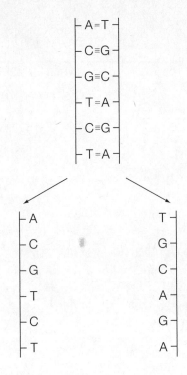

Figure 4.2 DNA replication

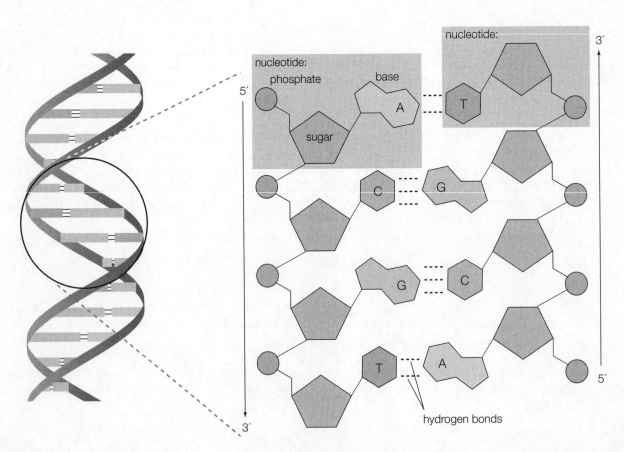

Figure 4.1 The structure of DNA

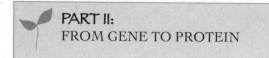

PART II:
FROM GENE TO PROTEIN

DNA ⇨ RNA ⇨ Protein

The diagram above summarizes the process used by the cell to make a protein: The gene's DNA is "transcribed" or copied into a strand of messenger RNA, which is then "translated" or used to synthesize the protein. Here are the details.

DNA is a linear molecule with a particular base sequence. A region of DNA that encodes for the structure of a particular protein is called a **gene**. **RNA**, or **ribonucleic acid**, is a linear, single-stranded nucleic acid like DNA *except* that RNA has **ribose** instead of deoxyribose, and it has the base **uracil** (U) instead of thymine. Uracil in RNA will hydrogen bond with adenine on another strand of nucleic acid. **Protein** is a linear molecule with a particular sequence of amino acids—usually several hundred.

Suppose gene *X* encodes for the structure of protein X. If a cell needs to synthesize protein X:

1. The double helix of gene *X* is "unzipped" through breakage of the hydrogen bonds.
2. A complementary strand of RNA called **messenger RNA (mRNA)** is synthesized along one of the gene's two DNA strands, called the template strand.
3. The synthesized mRNA detaches.
4. An organelle called a **ribosome** in the cell's cytoplasm attaches at one end of the mRNA and moves down its length. At the first AUG three-base sequence, or **codon**, the ribosome pauses and protein synthesis starts.
5. At the AUG codon, the amino acid methionine is brought in. The next codon on mRNA is the next three-base sequence; if it is CUU, it codes for the amino acid leucine, which is brought in and bonded to methionine. If the third codon is AGG, it codes for arginine, which is brought in and attached to leucine, and so forth. The **genetic code** has been determined and is shown in Table 4.1.
6. This process of reading the codon and attaching the specified amino acid to the previous one continues

First position in codon	Second position in codon				Third position in codon
	U	**C**	**A**	**G**	
U	Phe = phenylalanine	Ser = serine	Tyr = tyrosine	Cys = cysteine	**U**
	Phe = phenylalanine	Ser = serine	Tyr = tyrosine	Cys = cysteine	**C**
	Leu = leucine	Ser = serine	*stop* (UAA)	*stop* (UGA)	**A**
	Leu = leucine	Ser = serine	*stop* (UAG)	Trp = tryptophan	**G**
C	Leu = leucine	Pro = proline	His = histidine	Arg = arginine	**U**
	Leu = leucine	Pro = proline	His = histidine	Arg = arginine	**C**
	Leu = leucine	Pro = proline	Gln = glutamine	Arg = arginine	**A**
	Leu = leucine	Pro = proline	Gln = glutamine	Arg = arginine	**G**
A	Ile = isoleucine	Thr = threonine	Asn = asparagine	Ser = serine	**U**
	Ile = isoleucine	Thr = threonine	Asn = asparagine	Ser = serine	**C**
	Ile = isoleucine	Thr = threonine	Lys = lysine	Arg = arginine	**A**
	Met = methionine	Thr = threonine	Lys = lysine	Arg = arginine	**G**
G	Val = valine	Ala = alanine	Asp = aspartic acid	Gly = glycine	**U**
	Val = valine	Ala = alanine	Asp = aspartic acid	Gly = glycine	**C**
	Val = valine	Ala = alanine	Glu = glutamic acid	Gly = glycine	**A**
	Val = valine	Ala = alanine	Glu = glutamic acid	Gly = glycine	**G**

Table 4.1 The genetic code. This chart indicates the amino acid specified by each coding codon as well as the stop codons. This genetic code is "universal"—used by essentially all organisms.

until the ribosome reaches the codon UAA or UAG or UGA. These three codons do not code for amino acids; they are **stop codons**. When a stop codon is reached, protein synthesis ceases and the protein chain and ribosome detach from the mRNA.

Would the same protein be produced if the ribosome attached to the *other* end of the mRNA? No! How does the ribosome attach to the correct end? This is where a molecular detail becomes relevant. A strand of nucleic acid, be it DNA or RNA, has two ends that are different from each other. The end of the nucleic acid with the phosphate outermost is the 5' end, and the end of the nucleic acid with the bottom of the sugar outermost is the 3' end (see Figure 4.1 where the ends are labeled). The *ribosome attaches only to the 5' end of the mRNA* and moves down the mRNA from there.

Note that any double-stranded nucleic acid complex (the DNA double helix in Figure 4.1 or the DNA template strand and its temporarily bound mRNA strand) is "antiparallel" in that the two strands are oriented in opposite directions.

Note that while many genes are transcribed into messenger RNA for protein synthesis, other genes are transcribed into ribosomal RNA (rRNA) molecules, which are structural components of ribosomes. Still other genes are transcribed into transfer RNA (tRNA) molecules that function like taxis, each binding and transporting a particular kind of amino acid to the site of protein synthesis in the cell.

Practice 2:

On Worksheet 1, do the following steps only (refer to the example in Figure 4.3):

☐1. *Transcribe* the **original normal gene B** into the corresponding mRNA.

☐2. Then *translate* this mRNA into protein. Remember that the ribosome attaches to the 5' end, moves down the mRNA, starts protein synthesis at the first AUG, and terminates at the first stop codon. Ignore the shaded box in Worksheet 1.

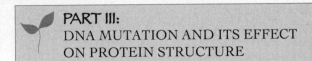

PART III:
DNA MUTATION AND ITS EFFECT ON PROTEIN STRUCTURE

A gene **mutation** consists of a change in the normal base sequence. The change can consist of a base **substitution**, which produces a different codon that may code for the *same* amino acid (**silent mutation**) or may code for a different amino acid (**missense mutation**) or may be a stop codon (**nonsense mutation**). A mutation may be an addition or deletion of one or more nucleotides and thus of one or more bases in the sequence (**frameshift mutation**). Since the code is read in threes, an addition or deletion of one or two bases will shift the reading frame for codons from that point onward. If *three* bases are added or deleted, then the reading frame is still in register—unaltered beyond the mutation site.

Practice 3:

Continue on Worksheet 1; work on your own or in pairs or as a group, as instructed.

For each **mutant** template strand given, do the following:

☐1. Compare the mutant *to the normal template* base sequence, and record whether there is a base substitution, addition, or deletion and how many bases are involved.

☐2. Transcribe the entire mutant template strand into mRNA.

☐3. Translate this mRNA into the protein (start at the first AUG codon; continue to the first stop codon).

☐4. Determine which amino acids are altered, how many amino acids there are, and whether this constitutes a silent, missense, nonsense, or frameshift mutation. *NOTE*: Ignore the shaded areas for now; Part IV of this lab explains the information requested in these shaded areas.

☐5. Check each other's results; discuss your results at your table and as a class as a whole.

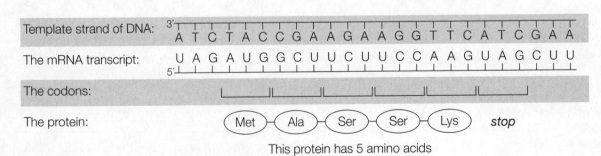

Template strand of DNA:	3' A T C T A C C G A A G A A G G T T C A T C G A A
The mRNA transcript:	U A G A U G G C U U C U U C C A A G U A G C U U 5'
The codons:	
The protein:	Met — Ala — Ser — Ser — Lys stop

This protein has 5 amino acids

Figure 4.3 Worksheet example

Transcribe, translate, and analyze the original normal gene *B*, and then do the same for its mutant forms.

Template DNA strand of *original normal gene B*:

3′ _____

C T C T A T A C G A A T C C C A T A G C C C T G G T A G A A T T C A C G A A

mRNA:

Amino acid sequence:

Number of amino acids in normal protein B: _____

Size & charge of each amino acid (refer to Table 4.2):

Template DNA strand of *mutant B1*:

Find and circle the mutation. This DNA change consists of: a substitution, addition, or deletion of _____ bases.

3′ _____

C T C T A T A C G A A T C C C A T A G C T C T G G T A G A A T T C A C G A A

mRNA:

Amino acid sequence (*circle any altered amino acids*):

This mutation is: silent, missense, nonsense, or frameshift.
Protein B1 has _____ amino acids; *this is: shorter, the same, or longer than normal.*

Size & charge of each amino acid (refer to Table 4.2):

Likely effect of mutation on protein folding and function?

Template DNA strand of *mutant B2*:

Find and circle the mutation. This DNA change consists of: a substitution, addition, or deletion of _____ bases.

3′ _____

C T C T A T A C G A A G C A C A T A G C C C T G G T T C G A T T C A C G A A

mRNA:

Amino acid sequence (*circle any altered amino acids*):

This mutation is: silent, missense, nonsense, or frameshift.
Protein B2 has _____ amino acids; *this is: shorter, the same, or longer than normal.*

Size & charge of each amino acid (refer to Table 4.2):

Likely effect of mutation on protein folding and function?

Template DNA strand of *original normal gene B* (for reference when working on this page):

3′ ───

C T C T A T A C G A A T C C C A T A G C C C T G G T A G A A T T C A C G A A

Template DNA strand of *mutant B3*:

Find and circle the mutation. This DNA change consists of: a substitution, addition, or deletion of _____ bases.

3′ ───

C T C T A T A C G A A T C C C A T A T C C C T G G T A G A A T T C A C G A A

mRNA:

Amino acid sequence (*circle any altered amino acids*):

This mutation is: silent, missense, nonsense, or frameshift.

Protein B3 has _____ amino acids; *this is: shorter, the same, or longer than normal.*

Size & charge of each amino acid (refer to Table 4.2):

Likely effect of mutation on protein folding and function?

Template DNA strand of *mutant B4*:

Find and circle the mutation. This DNA change consists of: a substitution, addition, or deletion of _____ bases.

3′ ───

C T C T A T A C G A A T C C A T A G C C C T G G T A G A A T T C A C G A A A

mRNA:

Amino acid sequence (*circle any altered amino acids*):

This mutation is: silent, missense, nonsense, or frameshift.

Protein B4 has _____ amino acids; *this is: shorter, the same, or longer than normal.*

Size & charge of each amino acid (refer to Table 4.2):

Likely effect of mutation on protein folding and function?

Template DNA strand of *mutant B5*:

Find and circle the mutation. This DNA change consists of: a substitution, addition, or deletion of _____ bases.

3′ ───

C T C T A T A C G A A T A T A G C C C T G G T A G A A T T C A C G A A

mRNA:

Amino acid sequence (*circle any altered amino acids*):

This mutation is: silent, missense, nonsense, or frameshift.

Protein B5 has _____ amino acids; *this is: shorter, the same, or longer than normal.*

Size & charge of each amino acid (refer to Table 4.2):

Likely effect of mutation on protein folding and function?

34

Template DNA strand of *original normal gene B* (for reference when working on this page):

3'
C T C T A T A C G A A T C C C A T A G C C C T G G T A G A A T T C A C G A A

Template DNA strand of *mutant B6*:
Find and circle the mutation. This DNA change consists of: a substitution, addition, or deletion of _____ bases.

3'
C T C T A T A C G A A T C C C A T A G C C C T C G T A G A A T T C A C G A A

mRNA:

Amino acid sequence (*circle any altered amino acids*):

This mutation is: silent, missense, nonsense, or frameshift.
Protein B6 has _____ amino acids; *this is: shorter, the same, or longer than normal.*

Size & charge of each amino acid (refer to Table 4.2):

Likely effect of mutation on protein folding and function?

PART IV:
DNA MUTATION AND ITS EFFECT ON PROTEIN FUNCTION (advanced)

The function of a protein is determined by its three-dimensional shape, which is determined by its amino acid composition. There are **20 different amino acids**. Each amino acid has an amine group and an acidic group. The amino acids differ from each other in the size and charge of the side group called the **R-group**. The R-group of some amino acids has an even distribution of its electrons and is called **nonpolar**. In some amino acids, the R-group has no net charge but the distribution of its electrons is not

even in some places, giving those places a slight negative or slight positive charge; such R-groups are called **polar**. The R-group of some amino acids has an extra electron, making the group **negatively charged**; the R-group of yet other amino acids has an electron missing, making the group **positively charged**.

The manner in which a protein chain folds into its final functional shape is affected by the nature and arrangement of the R groups of the amino acids in the protein chain. Remember that "opposites attract"—oppositely charged regions interact. Nonpolar amino acids have no charged regions and thus cannot interact with the polar water molecules; nonpolar amino acids are therefore hydrophobic and usually in the *interior* of the

Nonpolar amino acids = hydrophobic	Polar amino acids = hydrophilic	Positively charged amino acids = very hydrophilic	Negatively charged amino acids = very hydrophilic
Glycine	Serine	Lysine	Aspartic acid
Alanine	Threonine	Arginine	Glutamic acid
Valine	Cystine	Histidine	
Leucine	Asparagine		(Both of these have
Isoleucine	Glutamine	(All three of these have	mid-sized R-groups, with
Proline	Tyrosine	large R-groups.)	glutamic acid having the
Methionine			larger one.)
Phenylalanine			
Tryptophan			

Table 4.2 Charge on amino acids. Note that within a category, the amino acids are listed from small to large R-group, approximately.

folded protein structure, away from the surrounding water. Polar and charged amino acids interact with water and are generally found on the *surface* of the folded protein. Amino acid sequences with small R-groups can fold or coil *more compactly* than sequences with large R-groups. Proline has an unusual structure, producing a *bend* in the protein chain.

A protein's three-dimensional shape determines the function of the protein. When a gene is mutated and the amino acid sequence in the protein is altered, it may alter the protein's folded shape and, in turn, its ability to function normally.

Some mutant proteins are fragments, resulting from premature termination of protein synthesis due to a nonsense or frameshift mutation that produced a premature stop codon within the gene. Such fragments are generally nonfunctional unless the premature stop is very near the end and the missing few amino acids do not play a major role in determining the three-dimensional structure of the protein.

Practice 4:

☐1. On Worksheet 1, fill in the shaded areas: For the original normal protein and for each mutant, consult the table of amino acids (Table 4.2) and record the size and charge of the R-group of each amino acid in the protein chain. Use these abbreviations:

np = nonpolar	pol = polar
(+) = positive charge	(–) = negative charge
sm = small	mid = medium lg = large

☐2. Then describe the likely effect of the mutation on protein folding and function.

☐3. Finally, discuss the conclusions with students at your table and as a class.

Practice 5:

☐1. On Worksheet 2, create a gene! *NOTE*: The table of the genetic code (Table 4.1) is for codons on the **mRNA**, not the DNA; thus, first write a base sequence for the mRNA that encodes a protein—remember to include a start and a stop codon. Then write the base sequence of the corresponding DNA template strand.

☐2. Have your lab partner mutate the base sequence of the template DNA strand in some way.

☐3. Translate your original sequence; transcribe and translate the mutated sequence.

☐4. Evaluate the effect of this mutation on the structure and function of the protein.

☐5. Compare your results to those of others at your table; examine the variety of mutations and their effects on the protein product. Discuss the effects of the mutations and decide which mutation created at your table is likely to have the least effect and which will have the worst effect on protein function.

WORKSHEET 2: Create A Gene

Template DNA strand:

3′

Normal mRNA:

Normal protein:

Mutant mRNA:

Mutant protein:

Effect of the mutation on amino acid sequence:

Possible effects of the mutation on protein structure (folding):

PART V:
DNA MUTATION AND ITS EFFECT ON THE ORGANISM

A bacterial cell has one large, circular, double-helix DNA molecule as its genetic material. Nonbacteria are much more complex and thus have so much genetic material that it is more functional to have the complete set of genetic information subdivided into several linear pieces called **chromosomes**. Each chromosome contains one double-helix DNA and associated molecules.

Flowering plants (peas, snapdragons, etc.) and humans have two complete sets of chromosomes in each body cell. Thus, a cell contains *two* of each kind of gene—one gene present in each chromosome set. When the cell needs a particular gene's protein product, then both copies of the gene can be used for synthesis of mRNA and, in turn, protein.

What is the effect on the organism if one of the two copies of a gene is mutated and yields no functional protein? That depends on whether the cells of the organism can gear up and make the same quantity of protein from one copy that is normally made from two copies. If the cells can increase their rates of mRNA and protein synthesis in this way, then the organism will be normal. If the cells cannot gear up their rates of synthesis, then the deficiency in protein product will affect the organism's appearance or function. (Note that mutations differ in their effects: some mutations produce nonfunctional protein, while others produce abnormally functioning protein, while yet others produce no protein at all.)

Peas and snapdragons have a gene encoding a protein that is required for the production of red color pigment in flower petals. Pea and snapdragon plants that have *two normal copies* of the gene for red color will have red flowers.

Q1.	In peas, a plant with *one normal* copy of the gene for red color and *one mutant* copy, producing nonfunctional protein, also has red flowers. Explain fully why this plant's flowers are red.

Q2.	In snapdragons, if a plant has *one normal* copy of the gene for red color and *one mutant* copy, producing nonfunctional protein, its flowers are pink (i.e., pale red). Explain fully why this plant's flowers are pink rather than red.

Q3.	What color are flowers of peas and snapdragons that have both copies of this gene nonfunctional? _____

EXTENDED ASSIGNMENT

On another piece of paper, write a template DNA base sequence that encodes a protein with 10 amino acids. Transcribe it into mRNA and translate it into the encoded protein.

Q4.	How many of the bases in the DNA template strand could be substituted and still code for the *same* amino acid sequence?

Q5.	What *percentage* of the bases in your sequence could be mutated by base substitution resulting in silent mutations, thus still giving the same protein?

Q6.	You have the DNA base sequence of a particular gene from sweet pea and from species X and Y. A computer analysis reveals that species X has 95% DNA sequence homology to the pea gene (5% of the bases in the species X gene sequence are different), while species Y has 90% DNA sequence homology (10% of the bases in species Y gene sequence are different). Can you conclude that the *protein* in sweet pea is more similar in structure and function to that in species X than in species Y? Discuss.

4

LAB

MITOSIS
CELL DIVISION

I. Mitosis of a diploid cell
II. Mitosis of a haploid cell
III. When things go wrong . . .
IV. Real chromosomes!

5

OBJECTIVES

1. Be able to demonstrate with *labeled* chromosome models the process of mitotic cell division, in a diploid cell and in a haploid cell (Parts I and II).
2. Learn how a malfunction in mitosis can produce a polyploid cell, and subsequent normal divisions can lead to a polyploid plant (Part III.A).
3. Learn how a mutation can be perpetuated by mitotic cell division and result in a bud sport (Part III.B).
4. Learn to recognize cells under the microscope in the various phases of the mitotic cell cycle (Part IV.A).
5. Determine the proportion of the onion-root cell cycle devoted to mitosis and its phases (Part IV.B).

Why Bother with Having Mitosis?

If you are a **bacterial** cell, you don't need to go through the involved process of mitosis! A bacterial cell is quite simple, and thus not all that much genetic material is needed by the cell to operate. All of a bacterial cell's genetic information fits on a single large, circular DNA (double-helix shaped) molecule that is attached at one point to the inside of the cell's plasma membrane. When a bacterial cell divides, it first replicates its circular DNA molecule and then attaches the two copies at separate points on the inside of the plasma membrane; then the cell simply divides between the two points of attachment, producing two new cells— each containing one DNA molecule. Other than being attached at one point to the plasma membrane, the bacterial DNA molecule is free within the cellular fluid (cytoplasm), not confined within a nuclear envelope; thus, bacteria have no nucleus and are called **prokaryotes** (Greek: *pro* = before; *karyon* = nut, kernel and denotes the nucleus).

All non-prokaryotic cells have their genetic material confined within a **nucleus** and are called **eukaryotes** (Greek: *eu* = true). Eukaryotes are far more complex in their structure and function, and thus a eukaryote has about a thousand times more genetic material than a prokaryote does. It makes sense that a eukaryote does not have its genetic material in a single gigantic DNA molecule, but rather in several DNA segments; these DNA segments and their associated proteins are called **chromosomes**. Before a eukaryotic cell divides, all of its chromosomes are replicated (duplicated). It would then obviously not work to divide such a cell by simply running a partition down through its mass of replicated chromosomes. A precise mechanism is required to ensure that all of the replicated chromosomes assort precisely to form two identical new nuclei (and with the completion of cell division, two identical new cells). The mechanism for nuclear division is **mitosis** (Greek: *mitos* = thread; referring to the thread-like chromosomes that become visible during mitosis).

Each eukaryotic species has a characteristic number of chromosomes in its nucleus. Fungal cells have one set of genetic information (one copy of each kind of chromosome) in each nucleus and are called **haploid** (Greek: *haploos* = single); many plants and animals, including humans, have two sets of genetic information (two copies of each kind of chromosome) in each nucleus and are called **diploid** (Greek: *diploos* = twofold, double). When a haploid cell undergoes mitotic cell division, it results in two new haploid cells. When a diploid cell undergoes mitotic cell division, it results in two new diploid cells. Mitosis makes "more of the same"; a seedling "grows" into a mature plant by a combination of mitotic cell divisions and cell expansions.

Why do a lab with labeled chromosome models? There are two reasons. (1) Some students do not understand mitosis until they themselves actually move chromo-

NOTE: Your instructor will indicate the parts of this lab that your class will perform.

somes around on a desktop. (2) Quite a few students who can demonstrate mitosis nicely by drawing the process or by manipulating unlabeled pop-bead chromosome models will make major conceptual errors when asked to label their chromosomes with the alleles of genes. By working with these labeled chromosome models, students can fully learn mitosis—not just the chromosomal movements, but also the full meaning of DNA replication, homologues, and gene arrangement and assortment during mitosis.

BACKGROUND

As you work on the various parts of this lab, remember the following points:

- Mitotic cell division consists of nuclear division (**mitosis**) followed by cytoplasmic division (**cytokinesis**; Greek: *kytos* = cell, *kinesis* = motion).
- A **chromosome** contains a long **DNA** (double-helix shaped) molecule along with associated proteins.
- A **gene** is a section of the DNA that encodes a protein that affects a particular trait. For example, there is a *gene* for *flower color* in the garden pea. Many genes exist along the length of a chromosome's DNA.
- An **allele** is a particular version of the gene. For example, there is an *allele* for *red flowers* and another *allele* for *white flowers* for the flower color gene of the garden pea.
- A **diploid** cell (abbreviated 2*n*) has *two* sets of chromosomes. Two chromosomes that are of the same kind are called **homologous** (Greek: *homos* = same); the two **homologues** (also spelled "homologs") have the same array of genes along their lengths. Thus, a diploid cell has two of each kind of gene, one being on each homologue. For a particular gene, the two *alleles* present (one on each homologue) in a diploid cell may or may not be the same kind of allele.
- The **genotype** of a cell is the genetic makeup of the cell. By convention, the *same letter of the alphabet* is used as a symbol for the *alleles of one gene*. Here is the key to the allele symbols for the six genes that are on the chromosome models that you will use for this lab:

Gene for flower color:	Gene for plant height:	Gene for branching:
R = red *r* = white	*T* = tall *t* = short	*B* = branched *b* = single stem
Gene for leaf shape:	**Gene for leaf arrangement:**	**Gene for fragrance:**
E = elongated *e* = rounded	*A* = rosette *a* = along stem	*F* = fragrant *f* = not fragrant

- If you were examining only the gene for flower color and the gene for branching on the chromosomes in Figure 5.1, then the genotype would be written *RrBb*.
- The **phenotype** is the physical manifestation of the genotype. For the flower-color gene, allele *R* is for red and allele *r* is for white flowers. If a plant has one of each allele (genotype *Rr*) and has red flowers, then the "red" allele that determined the phenotype is called **dominant**; the "white" allele, whose effect is not visible, is called **recessive**. By convention, the dominant allele is symbolized by the *capital* letter and the recessive allele by the *lowercase* letter. In the key to the genes provided earlier, the capital letter indicates the dominant allele for each gene.
- A plant that is genotype *Rr* for flower color is said to be **heterozygous** for this gene. The genotype *RR* is called **homozygous dominant**, while *rr* is **homozygous recessive** (Greek: *heteros* = other, *zygotos* = yoked).
- The mitotic **cell cycle** (life cycle of a cell) consists of interphase, mitosis, and cytokinesis; during **interphase** the cell performs its normal function and also replicates its chromosomes by undergoing DNA replication in preparation for mitosis; after the nucleus divides by mitosis, the cytoplasm divides by cytokinesis.

PART I:
MITOSIS OF A DIPLOID CELL

The purpose of this section is to learn the process of normal mitotic cell division. Keep in mind that the molecular details and control mechanisms of mitosis are still largely unknown and are being actively researched.

MATERIALS

- red, navy, pink, and light-blue poster boards (a 22 in. × 28 in. board yields 15 pieces, 4.5 in. × 8.5 in.; 2 such pieces needed per model)
- sticky-backed 3⁄4-in. Velcro® or other brand of sticky-backed hook-and-loop fastener (12 squares per model)
- sticky-backed 3⁄4-in. round labels, 6 different non-black colors (4 labels of each color per model)
- sheet of felt (a quarter-section of a yard); if not felt, make sure Velcro adheres
- scissors, felt-tip pen, masking tape, and chalk

INVESTIGATION

Your instructor may first show a time-lapse video of mitosis (such as the *Living Cells* video by Pickett-Heaps and Pickett-Heaps, 1994).

☐1. Per group of four students, one student will make red chromosomes, one student will make navy-blue chromosomes, one will make pink chromosomes, and one will make light-blue chromosomes. (The four colors will be significant later on when you reuse these sets for the meiosis lab.) Each student will make two long, two intermediate, and two short chromosomes.

☐2. Use a 4.5-in. × 8.5-in. piece of poster board and Figure 5.1 as a guide to cut out and make the red or the pink poster-board chromosome models exactly as shown; use Figure 5.2 as a guide to cut out and make the navy or the light-blue poster-board chromosome models exactly as shown. The six types of shaded circles represent six different genes and will be six different colors of sticky-backed labels on your chromosome models.

NOTE: **Each gene is assigned a particular color that all students will use for all copies of that gene**; this system will be important later on.

Use a felt-tip pen to write the letters (allele symbols) on your gene labels as shown in the figures. Remember that many genes exist along the length of each chromosome; you are labeling only a few of them.

☐3. Stick a square of *fuzzy* Velcro onto the centromere region of the labeled side of each chromosome; onto the *unlabeled* underneath side of the chromosome, stick a square of *hooks* Velcro at the centromere region.

☐4. Attach your chromosomes, labeled side up, to your felt sheet to represent a diploid cell before division. See the diagram at the right. In Table 5.1, complete the genotype of your cell.

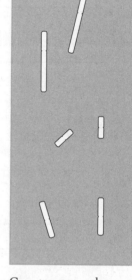

☐5. In preparation for **mitosis**, the chromosomes undergo **replication** during interphase. A replicated chromosome consists of two replicas that are genetically *totally identical* to each other and that *remain attached* to each other at one point, the **centromere**.

☐6. *Replicate your chromosomes*: Construct another set of chromosomes of the *same color*, exactly as shown in Figure 5.1 for the red or pink models and in Figure 5.2 for the navy or light-blue models.

☐7. To show the result of chromosomal replication, attach the replica chromosomes to the centromeres of the original chromosomes on the felt sheet, as shown at the right. Each replicated chromosome "condenses" into a compact structure that becomes visible under the light microscope during the first phase of mitosis, **prophase** (Greek: *pro* = before). Also, the nuclear envelope typically breaks down—except in fungi (an illustration of the "rule" in biology: "Never say never, never say always").

☐8. *Closely check your replicated chromosomes*. The two replicas of each chromosome must be totally identical—have exactly the same gene arrangement—and for a particular gene, the two replicas must have exactly the same allele. See the example at right:

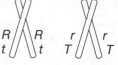

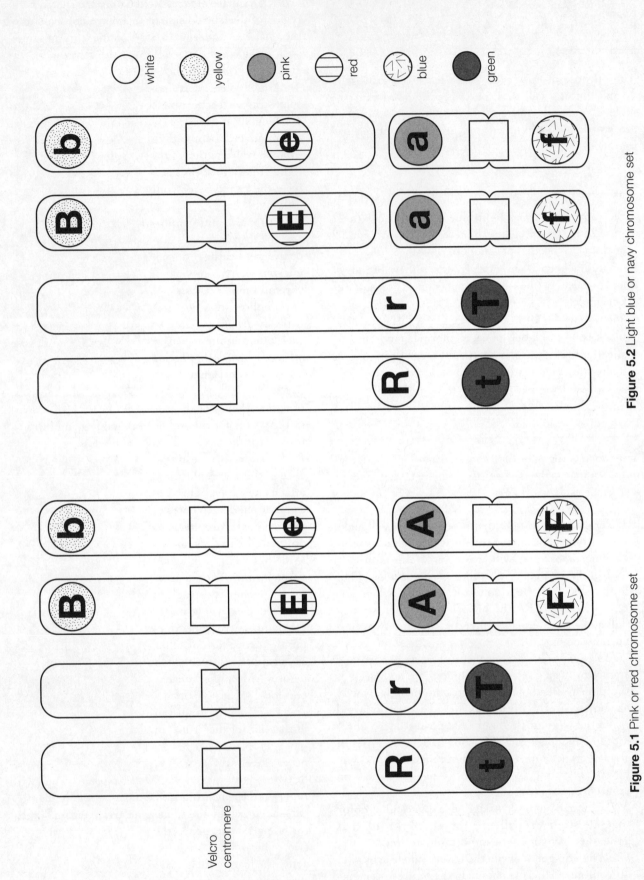

white
yellow
pink
red
blue
green

Velcro centromere

Figure 5.2 Light blue or navy chromosome set

Figure 5.1 Pink or red chromosome set

42

□9. Also in prophase, **spindle fibers** extend from the two regions called poles, attach to the centromeric region of each replicated chromosome, and then move each replicated chromosome to the spindle's equatorial plane to accomplish **metaphase** (Greek: *meta* = after).

Demonstrate metaphase by drawing (with chalk) the spindle fibers and positioning your replicated chromosomes on the equator.

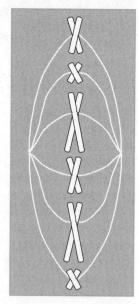

□10. *Demonstrate* **anaphase** (Greek: *ana* = away) by pulling the Velcro centromeres of the replicas apart and moving each of the two liberated replicas (now called chromosomes) to opposite poles. In a cell, the spindle fibers shorten as the chromosomes move to the poles.

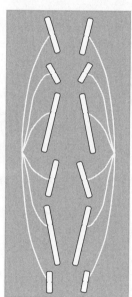

□11. *Demonstrate* **telophase** (Greek: *telo* = end), the phase when the chromosomes have reached the poles. Also, the spindle fibers disappear, the nuclear envelope reforms, and the chromosomes become less compact (decondense). Telophase is the fourth and last phase of mitosis (nuclear division).

□12. *Represent cytoplasmic division* (called **cytokinesis**, Greek: *kytos* = cell, *kinesis* = motion) to complete cell division by running a strip of masking tape down the middle of the felt sheet. Cytoplasmic division is accomplished in plants by the formation of a partition called the "cell plate" across the cell between the two new nuclei.

□13. Compare the chromosomal content of each of the two new cells (1 and 2) to each other and to the chromosomal content of the original cell. Record the chromosomal number and genetic makeup for the two new cells in Table 5.1.

□14. If both of your daughter cells are identical in genetic makeup (genotype) to your original diploid cell, then you have successfully performed mitosis! (If not, then try it again until you get it.)

5

	Chromosome number			Genetic makeup (genotype)
	# of long chromosomes	# of medium chromosomes	# of short chromosomes	
Original diploid cell before mitosis	2	2	2	*Rr Tt Bb Ee* _____ _____
New cell 1				
New cell 2				

Table 5.1 Mitosis of a diploid cell

43

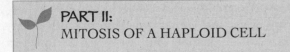

PART II:
MITOSIS OF A HAPLOID CELL

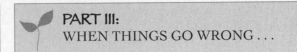

PART III:
WHEN THINGS GO WRONG . . .

The *plant* life cycle has a multicellular haploid structure alternating with a multicellular diploid structure; both of these multicellular structures develop via mitotic cell division. *Fungi* are eukaryotic haploids; they grow by mitotic cell division.

The purpose of this section is to show that the mechanism of mitosis works for a haploid cell as well as for a diploid cell, making in each case two new cells genetically identical to the original cell.

INVESTIGATION

☐1. Use the chromosome models of Part I. Attach to the felt sheet the chromosomes that represent the nuclear content of a *haploid* cell: one long chromosome, one medium chromosome, and one short chromosome. In Table 5.2 below, record the genotype of this "original haploid cell" that you have formed.

☐2. Replicate your chromosomes. Select, from your chromosome models, those that are replicas. Attach them to your chromosomes on the felt sheet.

☐3. Demonstrate prophase, metaphase, anaphase, telophase, and cytokinesis of mitotic cell division.

☐4. Record the chromosome number and the genetic makeup of the two new cells in Table 5.2.

☐5. If both of your two new cells are identical in genetic makeup (genotype) to your original haploid cell, then you have successfully performed mitosis! (If not, then try it again until you get it.)

III.A. WHAT HAPPENS WHEN SPINDLE FIBERS MALFUNCTION?

Have you ever flipped through a seed catalog such as Burpee's and noticed such flowers as the zinnia Big Tetra Mix, which are tetraploid zinnias having 6-inch flower heads? Did you wonder what that meant? Tetraploid plants may arise in several ways. The purpose of this section is to show how a defect in mitosis can produce a tetraploid.

INVESTIGATION

☐1. With your chromosome models, represent the chromosomal content of a diploid cell, replicate the chromosomes, and perform the process of mitosis up to metaphase.

☐2. Now start anaphase by separating the replicas of each replicated chromosome by pulling the Velcro centromeres apart. However, something goes wrong—the spindle fibers malfunction, the separated replicas (now independent chromosomes) fail to move to opposite poles, and there is no cytokinesis!

☐3. The chromosomes remain in one mass, around which a new nuclear envelope eventually forms.

☐4. What is the chromosomal composition of your cell now? Fill in Table 5.3.

	Chromosome number			Genetic makeup (genotype)
	# of long chromosomes	# of medium chromosomes	# of short chromosomes	
Original haploid cell before mitosis	1	1	1	
New cell 1				
New cell 2				

Table 5.2 Mitosis of a haploid cell

44

	Chromosome number		
	# of long chromo-somes	# of medium chromo-somes	# of short chromo-somes
Original diploid cell before mitosis	2	2	2
Resulting cell after failed anaphase of mitosis			

Table 5.3 Malfunction of spindle fibers during mitosis of a diploid cell

A cell that has two copies of each kind of chromosome is called **diploid**; a cell with four copies of each kind of chromosome is called **tetraploid** (Greek: *tetra* = four). Any cell that has more than two full sets of chromosomes is a type of **polyploid** (Greek: *poly* = many). What you have produced in this exercise is a tetraploid cell.

On rare occasions, this failure of cell division resulting in a polyploid plant happens in nature. Plant breeders and researchers can also induce polyploidy in the lab by exposing plant cells to the chemical colchicine, which interferes with the proper functioning of the spindle fibers.

Mitosis in a Tetraploid Cell

A cell—be it haploid or diploid or polyploid—can undergo the mechanism of normal mitosis and cytokinesis to produce two new cells that are genetically identical to each other and to that original cell. Thus, the tetraploid cell can undergo normal mitosis and cytokinesis, producing two new tetraploid cells. If the polyploidization event produced a tetraploid cell at the very beginning of embryonic development of a plant, subsequent *normal* mitotic cell divisions throughout development would produce a tetraploid plant.

Polyploid plants are healthy and often more vigorous than the diploid; remember the six-inch flower heads on the tetraploid zinnias? It is not understood yet why polyploid plants are bigger and better than diploid plants when most polyploid animals cannot develop normally and die.

III.B. WHAT HAPPENS WHEN THERE IS A MUTATION?

Occasionally you may find a plant with red flowers, except for one branch that has all white flowers on it. Several new apple varieties were discovered as "bud sports"—one branch bearing a different type of apple than what was on the rest of the tree. What happened? There has been a genetic change, a **mutation** (Latin: *mutare* = to change).

INVESTIGATION

☐1. Using your chromosome models and felt sheet, represent a diploid cell of genotype *Rr*. Suppose this cell is located at the growing tip of a shoot. Suppose this cell has undergone a mutation, a genetic change in the allele for red flowers so that it now specifies white flowers; thus, the diploid cell is no longer of genotype *Rr* but rather *rr* due to this mutation. Represent this mutation by replacing the *R* allele label with an *r* allele label on your chromosome model.

☐ 2. Now replicate your chromosomes. The two homologs of the long chromosome both now bear *r* alleles due to the mutation, and the replicas of each also bear *r* alleles for the flower-color gene.

☐3. Proceed with prophase, metaphase, anaphase, and telophase.

☐4. What is the genetic makeup of each of the two new cells produced by mitotic cell division from the mutated cell? Fill in Table 5.4.

Q1. If these two new cells now undergo normal mitotic cell division to grow into a new shoot, what flower color will be produced on this new shoot?_____

What flower color is on the rest of this plant?

_____

	Genotype of cell for flower color	Flower color specified by this genotype
Unmutated original diploid cell	*Rr*	red flowers
Diploid cell after mutation		
New cell 1		
New cell 2		

Table 5.4 Mitosis of a mutated diploid cell

5

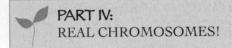

Chromosome models are great for learning the process of mitosis, but there's nothing like also seeing the chromosomes and observing the stages of mitosis in actual cells. The purpose of this section is to observe real chromosomes—in a chromosome squash that you make yourself, as well as in a prepared slide of a longitudinal section of an onion-root tip.

Remember that *chromosomes condense and become visible at the beginning of mitosis* and remain visible throughout mitosis; chromosomes are not visible during interphase (the nondividing portion of the cell cycle). To see real plant chromosomes, you must therefore look at dividing cells. In plants, dividing cells are not present throughout the body, as is the case for animals; rather, *plants have localized regions of cell division* called **meristems** (Greek: *meristos* = divided, divisible). There are **apical meristems** located at the tips of all the roots and shoots, and there is a thin layer of lateral meristem found in the rest of the plant body (in a tree, this layer is located between the wood and the bark). The most convenient region in which to see mitosis is in the apical meristem of an actively growing root.

IV.A. MAKE YOUR OWN PLANT CHROMOSOME "SQUASHES"

MATERIALS

- cuttings of *Zebrina pendula* (wandering jew), rooting in water and having new roots no longer than 1–2 cm
  - New *Zebrina* root tips are preferred because they are mitotically very active throughout the 24-hour cycle.
  - If *Zebrina* is unavailable, then onions from a farmer's market (i.e., unsprayed with root inhibitors) or spring flower bulbs will grow roots when positioned on top of water-filled, narrow jars, with the bulbs' basal discs immersed in water.
- 3 hot plates
- boiling stones
- 2 Pyrex® baking dishes
- distilled water
- metal test tube rack in each baking dish
- large flask, beaker, or pot (for extra hot water)
- autoclave gloves
- single-edge razor blades
- microscope slides
- cover slips
- paper towels
- dissecting needles
- dropper bottles of 1N HCl (this concentration available from Carolina or Ward's or other supplier)
- dropper bottle of TBO stain (0.5% toluidine blue O [aqueous]) (dilute the 1% solution available from Carolina or Ward's or other supplier)
- compound microscope, 1 per student
- lens paper
- clear fingernail polish (optional)
- protective gloves

> A NOTE ON SAFETY: Hydrochloric acid (HCl) is corrosive and toxic (HCl is the main ingredient in certain toilet bowl cleaners); toluidine blue O (TBO) is an irritant. Work in a well-ventilated area such as a fume hood. Avoid contact with skin and clothing; wear safety glasses and protective gloves. See the supplier's "Material Safety Data Sheets" for more details.

INVESTIGATION

This method was developed by Joseph J. Nickolas, modified by Connie Roderick, and modified by Glen G. Wurst.

NOTE:
- The warm acid treatment kills, preserves, and softens the tissue of the root tip.
- The TBO enters the cells and stains the chromosomes.
- **Important: The timing in this procedure must be exact!** Thus, you must fully read and understand the procedure *before* you start to work.

☐1. In a well-ventilated area such as in a fume hood, put a Pyrex® baking dish containing a metal test tube rack onto each hot plate, add several centimeters of distilled water, and bring to a *gentle rolling boil*; a few boiling stones may help; have more hot distilled water available in a large flask, beaker, or pot on another hotplate, to be used to maintain the water level in the baking dishes throughout the procedure. *CAUTION*: Use autoclave gloves when handling hot objects.

☐2. Select a new *Zebrina* root that is 1- to 2-cm long; cut it off with a razor blade. *Be sure to remove the whole root.* (Results are not as good when using longer roots.) Return the *Zebrina* to the water.

☐3. Place the root onto a microscope slide; cut off and *keep* 1 cm *of the tip*; discard the rest of the root.

☐4. Position the 1-cm root tip in the center of a microscope slide, and quickly add 4 drops of 1N HCl (hydrochloric acid).

☐5. Immediately place the slide onto the rack above the gently boiling water, and leave it there for **exactly 1-1/2 minutes**. During this heat treatment, use a dissecting needle to gently roll the root tip back and forth in the acid; add more acid if needed to prevent it from drying up.

☐6. Remove the slide from the rack; quickly place this slide onto folded paper towels and use the towel to blot the acid from around the root tip.

☐7. To the blotted root tip, quickly add about 3 drops of TBO stain, or enough to immerse the root tip in it.

☐8. Immediately place the slide onto the rack above the gently boiling water again, and leave it there for **exactly 1-1/2 minutes**. During this heat treatment, use a dissecting needle to gently roll the root tip back and forth in the TBO stain; add more stain if needed to prevent it from drying up.

☐9. Remove the slide from the rack; place this slide on folded paper towels and blot the remaining stain from around the root tip. Immediately add only 1 drop of TBO stain. Use a dissecting needle to center and slightly flatten the root tip. Add a clean cover slip.

☐10. Place this slide with root tip and cover slip onto several layers of paper towel on a hard, flat surface; fold the paper towels over the cover slip area; position your thumb onto the paper towels directly over the cover slip and press down *hard* on the cover slip, *without any twisting*. You may roll your thumb from side to side, but do not twist it. Twisting will bunch up the cells. You now have a "squash mount slide." Blot any liquid that has been forced out from under the cover slip.

☐11. Remove the squash mount slide from the paper towels and place this slide onto the rack above the gently boiling water, for **exactly 1 minute**. This stain-clearing step improves the contrast between the stained and unstained regions of the root tissue.

☐12. Remove the squash mount slide, dry the bottom of the slide, and view the slide under a compound microscope—use low power first, and then move to higher power to see more detail. See Lab 2 (Microscopy and Plant Cells) for instructions on using a microscope.

☐13. Find a region where cells contain dark-blue nuclei or dark-blue strands (chromosomes). Identify cells that are in interphase, prophase, metaphase, anaphase, and telophase and draw them in the space provided. If your slide is not very good, then get another root and repeat the procedure exactly.

☐14. If you have a good temporary wet mount, you may wish to preserve it by sealing the edges of the cover slip with clear fingernail polish.

Interphase
Prophase
Metaphase
Anaphase
Telophase

Q2. Are any of the cells ruptured and their chromosomes spread out? _____
If so, then count the number of chromosomes for several cells; note that you may get a faulty count if two chromosomes lie on top of each other. How would the count of a ruptured metaphase cell compare to that of a ruptured anaphase cell?

Based on your observations, the diploid (2n) chromosome number per cell for your plant is _____.
Consult the Index to Plant Chromosome Numbers (IPCN) on the website of the Missouri Botanical Gardens at http://mobot.mobot.org/W3T/search/ipcn.html and find the diploid (2n) chromosome number for your plant (IPCN lists plants by family; *Zebrina pendula* is in the family Commelinaceae, and onion is *Allium cepa* in the family Liliaceae). (Note that website addresses may change; if you cannot locate a website by a given address, then do a web search for the organization and/or the name of the database.) According to the IPCN, the 2n chromosome number per cell for your plant is _____.

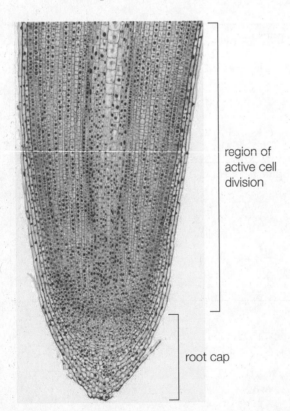

region of active cell division

root cap

Figure 5.3 Onion-root tip (longitudinal-section)

IV.B. HOW LONG DO THE PHASES OF MITOSIS LAST IN ONION-ROOT TIPS?

A root grows in length because there is a region of active mitotic cell division and cell elongation just behind the root cap at the tip of each root.

The purpose of this part is to locate the meristem (region of active mitosis) in a longitudinal section of an onion root, and to observe in this region the mitotic phases and determine their duration in the cell cycle. The **cell cycle** consists of interphase (during which time the cell performs its normal function as well as replicates its chromosomes in preparation for division) followed by mitosis and cytokinesis.

MATERIALS

- onion (*Allium*) mitosis, root tip, l.s., microscope slides (Carolina HT-30-2396; Ward's 91 W 7040)
- compound microscope, 1 per student
- lens paper

INVESTIGATION

☐1. Scan the slide of the onion-root tip at low power and compare it to the section shown in Figure 5.3. Find the root cap.

☐2. Switch to the next higher power and locate the region of active mitosis behind the root cap. You will see that some of the cells in this region have clumps of stained strands (chromosomes) in them.

☐3. Switch to the highest power to see the greatest detail. Evaluate several dozen cells until you feel confident in diagnosing the stage of the cell cycle. (refer to Figure 5.4):
- **interphase** (uniformly granular nucleus)
- **prophase** (nucleus with a random tangle of distinct strands)
- **metaphase** (chromosome strands arranged at the equatorial plane of the spindle)
- **anaphase** (two groups of rather distinct chromosome strands in the cell)
- **telophase** (two compact clumps of chromosomes far apart in the cell, with the beginning of the cell plate forming between the clumps to divide the cell).

☐4. Determine the relative length of each phase in the cell cycle by determining how many cells in your field of view are in each of the phases. To do this:
a. Locate the region of active cell division just behind the root cap.

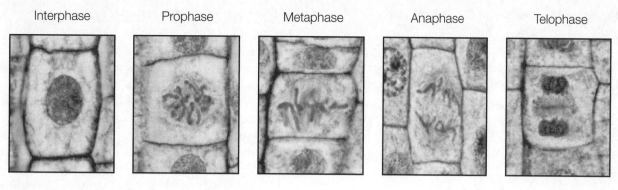

Interphase Prophase Metaphase Anaphase Telophase

Figure 5.4 The phases of mitosis in onion-root tip

b. Then *systematically go along rows of cells*, determining the phase of each cell in the row (without skipping any cells) until you have examined 20 cells.

Tally your results in Table 5.5; also record your totals in a table on the classroom board in order to come up with a class total for each phase.

□5. Calculate the percentage of time that a cell (in this region of the onion-root tip) spends in each phase of the cell cycle using your data and then also using the class data:

$$\text{\% of time that a cell spends in this phase} = \frac{\text{\# of cells in this phase}}{\text{total \# of cells examined}} \times 100\%$$

Q3. The *percentage* of the cell cycle that is spent in mitosis is called the **mitotic index**. This index can be calculated by determining the number of cells *within an active, mitotically dividing region* (meristem) that are in some phase of mitosis (prophase, metaphase, anaphase, or telophase):

$$\text{mitotic index} = \frac{\text{\# of cells in mitosis}}{\text{total \# of cells examined in the meristem}} \times 100$$

Calculate the mitotic index for onion based on your slide:

Calculate the mitotic index for onion based on the class data:

Q4. Are you more confident in the accuracy of the percentage calculated from *your* data or from the *class* data? Why?

Q5. The cell cycle of onion-root-tip cells lasts about 16 hours. How long, in terms of hours and minutes, does *mitosis* last in an onion-root-tip cell?

Phase of cell cycle	Number of cells *you* found in this phase	% of time the cell is in this phase (based on *your* data)	Number of cells *the class* found in this phase	% of time the cell is in this phase (based on *class* data)
Interphase				
Prophase				
Metaphase				
Anaphase				
Telophase				
Number of cells examined	Total = 20		Total =	

Table 5.5 Data sheet for onion mitosis

Questions To Contemplate And Discuss

1. In the squash mount slides (Part IV.A) and the prepared slides (Part IV.B), the chromosomes in the root-tip cells do not move. Why? Would you see them move if you took time-lapse photos of them over the next several hours? (Hint: Look at the chromosome-staining procedure used in Part IV.A of this lab; contemplate the effects that the treatments would have on a living cell.)

2. How would the mitotic index of a less frequently dividing region of a root differ from the index of an actively dividing meristematic region?

3. To determine the toxicity of chemicals and pesticides on plants, researchers investigate various plant characteristics, including the mitotic index of the chemically treated root tips as compared to that of untreated root tips. Explain what the mitotic index values would tell the researchers. If there was a reduced mitotic index in the root tips, what would that suggest about the plant as a whole?

ACKNOWLEDGMENTS

The chromosome-staining procedure was first presented by Joseph J. Nickolas (now Professor Emeritus, Northland Pioneer College) at the National Association of Biology Teachers (NABT) Convention, University of Nevada, Las Vegas, on October 24, 1981, in a workshop entitled "An Incredibly Fast and Simple Procedure for Making Squash Mount Slides of Plant Cell Chromosomes." Modifications were made by Connie Roderick (while at University of Wisconsin-Baraboo/Sauk County; presently at USGS, National Wildlife Health Center, Madison, WI) and further by Glen G. Wurst (Allegheny College). Used with permission.

LITERATURE CITED

Pickett-Heaps, J. D., and J. Pickett-Heaps. 1994. *Living Cells: Structure and Diversity* [videocassette]. Sunderland, MA: Sinauer.

Missouri Botanical Garden. 2006. "Index to Plant Chromosome Numbers (IPCN) Data Base," http://mobot.mobot.org/W3T/Search/ipcn.html (accessed February 19, 2006).

LAB 6 | SEED GERMINATION

I. What is inside a seed?
II. Dormancy: these seeds won't germinate; what do I do?
III. How is seed germination affected by environmental factors?

NOTE: The objectives are given at the beginning of each part of this lab.

BACKGROUND: WHERE AND WHAT ARE SEEDS?

Dispersal to other locations is critical for long-term survival of a species, since environmental conditions may change and kill the individuals in a particular location. Although plants are generally anchored to the ground or other substrate, they are capable of dispersal at some point in the life cycle. Plants have evolved two means of dispersal: spores or seeds.

The **seedless plants** disperse as **spores**. Spores are thick-walled single cells that are typically wind dispersed. The seedless plants include the bryophytes (liverworts, hornworts, mosses), the lycophytes (club mosses, quillworts, *Selaginella*), *Psilotum*, horsetails, and ferns.

The **seed plants** disperse as **seeds**. Each seed includes an embryonic plant supplied with stored food. The seeds are dispersed by various means, depending on the design of the seed.

The seed plants are divided into two groups:

1. The **gymnosperms** are the naked-seeded plants that include the conifers, ginkgo, cycads, and several others (Greek: *gymnos* = naked; *sperma* = seed).
2. The **angiosperms** are the flowering plants (Greek: *angeion* = container; the seeds are contained, enclosed within a fruit, although at maturity the fruit may split open to release the seeds).

All seed plants produce **ovules** and **pollen** (see Figure 6.1). Embedded within the tissue of each ovule is an egg cell ("ovule" is diminutive of the Latin: *ovum* = egg); the egg cell is located near a tiny opening in the ovule called the micropyle (Greek: *mikros* = small; *pyle* = gate). The pollen grains contain the **sperm** cells

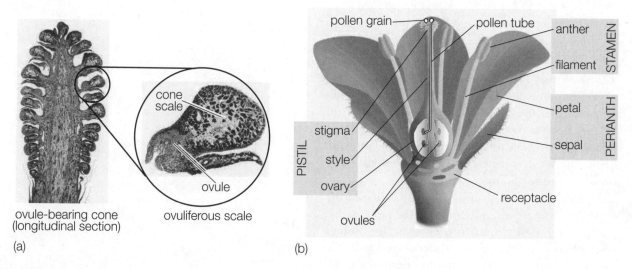

Figure 6.1 Location of ovules in seed plants. (a) Pine: ovules on the scales of young female cone; pollen is produced by separate male cones. (b) Flower: ovules within the ovary; pollen produced by anthers.

NOTE: Your instructor may have the class do only certain parts of this lab or may have different student groups do different parts and then report their results to the class.

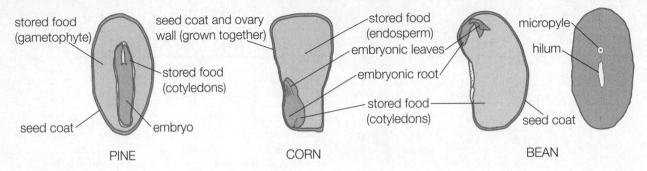

Figure 6.2 Seed structure

(Latin: *pollen* = fine flour). After a sperm fertilizes the egg cell in an ovule, the ovule matures into a seed.

- In **gymnosperms**, the seed is *naked* because the ovule from which it developed was naked (not enclosed in tissue). In pine, for example, two ovules lie *on top of* each cone scale of a female cone (Figure 6.1a), while many pollen grains are produced by a small, soft, short-lived male pollen-bearing cone. A pollen grain that is blown to the ovule's micropyle opening will germinate and grow into the ovule in order to deliver the sperm to one of the 2–3 egg cells within the ovule. After an egg is fertilized, the ovule develops into a naked pine seed on top of a cone scale.

- In **angiosperms**, the seed is *enclosed* because the ovule from which it developed was enclosed within the ovary of the flower's pistil (Figure 6.1b). A pollen grain that lands on the pistil's stigma will germinate and grow a long pollen tube through the style, into the ovary, and through the micropyle opening into the ovule, to deliver the sperm to the single egg cell within the ovule. After the egg is fertilized, the ovule develops into a seed and the surrounding ovary wall develops into fruit tissue.

A **seed** has the following parts (see Figure 6.2):

- The **seed coat** is a protective covering that developed from the outer layers of the ovule and is thus material from the mother plant. The seed coat may have structural modifications to assist in seed dispersal; for example, many pines produce "winged" seeds, the wing being a dry extension of the seed coat that aids in gliding through the air.

- The **embryo** is the offspring plant that developed from the fertilized egg. Attached to the embryo are one or more fleshy, nutrient-rich structures called **cotyledons** (Greek: *kotyledon* = cup-shaped hollow). Gymnosperms tend to have several cotyledons per embryo. The angiosperms are traditionally divided into two groups: the **monocotyledons** or monocots (one cotyledon per embryo) and the **dicotyledons** or dicots (two cotyledons per embryo).

- The **nutritive tissue** (stored food) in the seed provides the germinating seedling with energy for a brief period; if the seedling reaches light before the stored food is used up, the seedling can begin making its own food by photosynthesis. The nutritive tissue in gymnosperm seeds has a different developmental origin than that in angiosperm seeds (see your textbook for details). The nutritive tissue in angiosperm seeds is called the **endosperm** (Greek: *endon* = within; *sperma* = seed); in some angiosperm species, the endosperm is absent in mature seeds since it is fully absorbed into the embryo's large fleshy cotyledon(s) during seed development.

A plant reproducing by seed is somewhat analogous to a mother putting her child on a plane with a bag lunch and sending him out into the world.

The purpose of this lab is to investigate seeds and seed germination. You will examine the structures in various seeds, learn about the germination process and dormancy, and investigate the effect of various environmental factors (light, salt, acid, temperature) on the germination rate.

PART I:
WHAT IS INSIDE A SEED?

OBJECTIVES

1. Locate and compare the different parts of the seed of a gymnosperm, an angiosperm monocot, and an angiosperm dicot.
2. Understand the function of each part of the seed.

MATERIALS

* gymnosperm: pine seeds (can be untoasted pine nuts, from a grocery or health-food store)
* monocot: fresh corn cut from cob, and germinating corn with embryo enlarged and barely protruding
* dicot: bean seeds, soaked to hydrate (from the grocery store: kidney, red, or other large beans that show the micropyle well)
* single-edge razor blades
* dissecting needles
* dissecting microscope, 1 per student pair

INVESTIGATIONS

☐1. Dissect each of the seeds provided.

☐2. *For each seed:* Find the protective **seed coat** and peel it off. Note that in store-bought pine nuts, the hard seed coat has already been removed. Note that in *corn* and other members of the grass family, what appears to be the seed coat on the kernel or grain is actually the *seed coat and ovary wall grown together into a single layer*; the corn kernel is actually a fruit.

☐3. Refer to Figure 6.2 and find the **embryo**, which is a tiny plant in a resting state; when conditions are right for germination, the embryo will resume growth.
* To find the embryonic *pine* tree in the seed, cut the pine seed in half lengthwise.
* At the bottom on one side of the peeled *corn* kernel, find the waxy nub that contains the embryo; the embryo is enlarged and easier to see on the germinating corn; cut the embryo vertically.
* To see the *bean* embryo, gently pry the two halves of the bean apart.

☐4. Examine each seed's embryo under the dissecting microscope and find:
* **embryonic root** or **radicle** (Latin: *radix* = root)
* **embryonic shoot** or **plumule** (Latin: *plumula*, diminutive of *pluma* = feather) with embryonic leaves that sometimes look like tiny feathers.

* fleshy **cotyledon(s)**

Which of your specimens are *dicots*? _____
_____

☐5. Find the **nutritive tissue** enclosed within the seed but not part of the embryo and its cotyledon(s). Which seeds had their stored nutrients present *only* in their large, fleshy cotyledons? _____
_____

☐6. Look at the edge of a large, undissected bean seed (see Figure 6.2). Find the seed's "belly button" called the **hilum** (Latin: *hilum* = a little thing); this is the point at which the seed was attached to the inside of the ovary. Note that the hilum is not always so little; the large white mark on a buckeye or horse-chestnut seed is the hilum!

☐7. Adjacent to the hilum on the bean seed, find a tiny bump called the **micropyle** (Greek: *mikros* = small; *pyle* = gate); see Figure 6.2. The micropyle was the small opening or "gate" through which the pollen tube entered the ovule in order to deliver the sperm to the egg—resulting in fertilization, which triggered the development of the ovule into the seed that you are examining.

Q1.   Botanically speaking, what is the thin "hull" that gets stuck between your teeth when you eat corn?

Q2.   Botanically speaking, what are those little waxy nubs that squeeze out of corn kernels and are left on your plate after eating corn?

Q3.   Botanically speaking, what are the little fleshy comma-shaped things that float around in bean soup?

6

PART II:
DORMANCY: THESE SEEDS WON'T GERMINATE; WHAT DO I DO?
(duration: may last entire term)

OBJECTIVES

1. Understand the difference between seed quiescence and seed dormancy.
2. Understand the methods by which seed dormancy can be broken in various species.
3. Practice the scientific method by formulating a hypothesis and properly designing and conducting a long-term experiment testing your hypothesis.

BACKGROUND

Some seeds, such as those of many domesticated plants that you buy in seed packets, are simply **quiescent** when released from the plant; they are physiologically ready to germinate as soon as you provide them with appropriate temperature, water, light conditions, and oxygen levels. Some plant species release seeds that are **dormant**; these seeds are not physically or physiologically ready to germinate even if you provide them with appropriate conditions.

Physical dormancy can be due to a thick seed coat that must first be worn away to some degree, such as by rolling around on the ground or by passing through the digestive tract of a bird or other animal. People mimic this process by **scarification**—rubbing seeds between sandpaper or soaking them in acid.

Physiological dormancy requires a period of "after-ripening," during which time biochemical changes occur that bring the seed to physiological readiness for germination. Physiologically dormant seeds of plants in temperate regions often require exposure to a cold period (winter) for a certain length of time to trigger the after-ripening processes. People mimic this process by **stratification**—placing seeds between moist layers of sand or paper towels in a refrigerator. Physiologically dormant seeds of some desert plants require several rains to leach out inhibitory chemicals that have prevented germination. Many dormant seeds from fire-prone, as well as fire-free, ecosystems will germinate when exposed to plant-derived smoke and other fire-related factors; this has become a very active area of research in recent years.

Q4. What adaptive advantage would there be for each of these methods of breaking dormancy?

scarification of thick seed coat:

cold period:

several rains:

fires:

MATERIALS

Collect one of the following types of seeds:
• Seeds of a native plant species of your area
• Seeds of apple or another temperate region fruit (everyone can eat one variety of apples and save the seeds)
• Seeds of orange or another tropical or subtropical region fruit (everyone can eat one variety of orange and save the seeds)

INVESTIGATIONS

(It's best to start these early in your term!)

Design a series of experiments to determine the procedure for successful seed germination for your chosen plant species. This investigation can proceed until the end of the term. (The class might work as a whole, dividing the work among the various student groups.)

54

☐1. Decide on the seeds to be used and the question to be investigated.
  a. Consider whether seed dormancy in the plant's native environment would be advantageous, and what factors might be involved in breaking it. For example, it is optimal for seeds of desert plants to germinate at the start of the wet season, and for seeds of temperate plants to germinate after the cold winter rather than on a balmy day in mid-winter.
  b. Evaluate the thickness of the seed coat on your seeds; could that be a factor?
  c. Do a literature search on your plant or related species; this might give you some insights.
☐2. Formulate your question and your hypothesis (what you think will happen), and record them on the data sheet for Part II.
☐3. Determine the details of your experimental design.
  a. Determine the total number of seeds needed, the number of seed groups, the number of seeds in each group, the manner in which each seed group will be treated.
  b. Will you use soil in pots or moist paper towels in a petri dish or plastic bag?
  c. Should you use sterile materials and surface-sterilize the seeds to eliminate mold? Would it matter? You could find out by having one student group use sterile dishes, paper towels, forceps, water, and surface-sterilized seeds and another group use regular materials and seeds. (Seeds can be surface-sterilized by putting them in 10% bleach solution for 5 minutes and then rinsing them thoroughly with water for 5 minutes. *NOTE: Use safety glasses. Bleach is corrosive; do not get in eyes, on skin or on clothing.*)
  d. Remember that the hallmarks of good experimental design are *replication*, a *control*, and a *single variable*.
☐4. Present your experimental design to your class for discussion and suggestions; after the discussion, revise your experimental design.
☐5. Determine all the materials that you will need; acquire the materials and set up the experiment.
☐6. At the end of the term, compile your data, and summarize it in graphical form. Prepare a written or oral report, as instructed, presenting and discussing your results. Compare your results to the information that you found in your literature search. Discuss the adaptive advantage of the method for breaking seed dormancy that you determined through experimentation for this species. If various kinds of seeds were investigated by your class, then discuss how the results compare.

PART III:

HOW IS SEED GERMINATION AFFECTED BY ENVIRONMENTAL FACTORS?
(duration: 2–3 weeks depending on seeds used)

6

OBJECTIVES

1. Understand the basics of the seed germination process.
2. Practice the scientific method by formulating a hypothesis and properly designing and performing an experiment.
3. Investigate one or more of the following:
   • Effect of various temperatures on the germination rate of seeds
   • Effect of light vs. darkness on the germination of the seeds of different species
   • Effect of environmental pollutants (various concentrations of salt or of acid) on the germination rate of seeds

BACKGROUND

The environmental influences on—as well as the initiation, biochemical processes, and genetic control of—seed germination are very active areas of research. These research findings will undoubtedly have agricultural applications.

Mature seeds are very dry, containing only 5–15% of water by weight. The first step in the germination process is the uptake of water, called **imbibition**. Imbibition is a passive process that swells the seeds—whether or not the seed is viable. In a viable seed, imbibition activates the enzymes; some enzymes digest starch into sugars, and others digest storage proteins into amino acids. These materials are then broken down further by the enzymatic process called **cellular respiration**, releasing energy that the seed's embryo will use to resume growth. The rest of the enzymes and other hydrated molecules engage in a complex cascade of biochemical processes that result in the development of the embryo into a seedling and an independent plant.

Q5. If a seed is buried deeply, and its germinating seedling has used up its stored nutrients before it could reach the light, what is the consequence?

55

DATA SHEET FOR PART II: Seed Dormancy

Your question: _____

Your hypothesis: _____

At each time point, record the *number of germinated seeds* and calculate the *% germination* in each treatment.

Length of treatment or time since treatment	Description of treatments, and number of seeds in each treatment									
	# germ	% germ	# germ	% germ	# germ	% germ	# germ	% germ	# germ	% germ
t= 0 days										
t=										
t=										
t=										
t=										
t=										
t=										
t=										
t=										
t=										
t=										
t=										
t=										
t=										

Seeds need more than just water for successful germination. Consider the following:
- You may have noticed the different effects that a mild spring and a cold spring have on the germination rate of flower and vegetable seeds in your garden. People make comments such as "spring is late this year." How is seed germination affected by **temperature**? Are the seeds of all species affected to the same degree? Do some seeds prefer a cooler temperature than others do?

- While hiking in a forest, you may have noticed the lush growth of seedlings in an opening created by a fallen tree—seedlings that are not growing in the surrounding shady forest. Did the sudden exposure to **light** trigger these seeds to germinate, or did they germinate everywhere but survive only in the sunny opening? Do all seeds require light, or do some kinds require darkness, and other kinds germinate regardless of the light condition? Is 24 hours of light better than a day/night cycle or total darkness?

- The effect of acid rain on the health of our vegetation is of great concern. How does exposure to **acidity** affect seed germination? What concentration of acid permits germination? Are the seeds of all species affected to the same degree by acidity?
- Salt runoff from roadways and driveways may damage adjacent vegetation. There are various kinds of salts on the market in addition to rock salt (NaCl). Are all the different **salts** equally damaging to the germination of seeds of one particular species? How concentrated does the salt solution have to be to prevent seed germination in various species?

The purpose of this part of the lab is for you to design and conduct an experiment to test your own hypothesis about the effect of a particular environmental factor on seed germination. If different student groups investigate different factors and then report on the results, you will learn about the effect of various environmental factors on seed germination.

MATERIALS

- paper towels
- petri dishes (several sleeves)
- Parafilm M®
- scissors
- graduated pipets and pipet pumps

To investigate temperature:
- areas/incubators/growth chambers warmer and cooler than the lab room's temperature
  (seed catalogs such as Burpee and Park's Seeds sell electric "seedling heat mats")
- various kinds of seeds (packets of vegetable or flower seeds; grocery store seeds—bags of dried beans, peas, and lentils or jars of herb seeds such as dill, celery, etc.)
- distilled water

To investigate light:
- bench by the window or under lights
- dark: in a box or light-tight cabinet
- seeds of *Coreopsis* (tickseed)
- seeds of *Oenothera* (evening primrose)
- lettuce seeds: looseleaf, Grand Rapids, and other varieties
- radish and other seeds
- distilled water

To investigate acidity:
- vinegar (this is diluted acetic acid, 4–8% usually)
- radish and other kinds of seeds
- distilled water
- graduated cylinders, graduated pipets with pipet pumps, and beakers for making dilutions

To investigate salts:
- students and staff bring in various brands and types of road salts
  (*NOTE*: Some kinds of salt crystals are very hygroscopic, absorbing much moisture from the air; within a few days, a paper bag of such crystals will be sitting in a pool of liquid. Therefore, keep your salts in *sealed plastic bags or jars* until use.)
- radish and other kinds of seeds
- distilled water
- balances, weighing paper, and spatulas
- graduated cylinders and beakers for making solutions

NOTE: Radish seeds germinate within days under appropriate conditions.

INVESTIGATIONS

☐1. Working in groups, pick one environmental factor; each student group can pick a different factor to test. You might even decide to have the class as a whole do one large experiment investigating the same factor, with the data from all the groups being pooled for analysis at the end.

☐2. Formulate your question and your hypothesis (what you think will happen), and record them on the data sheet for Part III.

☐3. Design an experiment to test your hypothesis, indicating the number of seed groups, the number of seeds in each group, the condition in which each group will be placed, the concentration of the solutions that you will mix, and so forth. Remember that the hallmarks of good experimental design are *replication*, a *control*, and a *single variable*. Select at least three different conditions for your variable; for example: at least three different temperatures or at least three different vinegar dilutions.

☐4. Present your experimental designs to the class, for discussion and suggestions. After the discussion, revise your experimental design, gather the materials, and set up the experiment.

☐5. *To keep your seeds constantly moist*: Fold up a paper towel into a square and cut off the corners so that the towel fits into a petri dish. This will serve as a wet platform that sits in a pool of solution in the dish and wicks up the solution to keep the seeds on top uniformly moist throughout the experiment.

☐6. *To prevent evaporation*: Seal the edges of each closed dish with a strip of Parafilm 1–1.5 cm wide. (To do this, hold the beginning of the Parafilm strip against the edge of the dish while pulling the strip with the other hand to stretch it and wrap it one and a quarter times around the dish; anchor the terminal stretched Parafilm end by pressing it onto the initial stretched

Parafilm on the dish—upon stretching, Parafilm becomes tacky and will stick to itself.)

☐7. *Vinegar dilutions* with distilled water can be expressed as a percentage (ml of vinegar in 100 ml of final solution). If a pH meter is available, you can determine the pH of all of your solutions. (The pH scale goes from 0 to 14, with 7 being neutral; the lower the pH value, the more acidic the solution; the higher the pH, the more alkaline.)

☐8. *Salt solutions* in distilled water can be expressed as a percentage (grams of salt per 100 ml of solution).

☐9. *To use sterile conditions*: All materials would need to be sterile—the dishes, paper towels, solutions, and seeds (to surface-sterilize seeds, see instructions in Part II).

☐10. Ideally, you should check your seeds for germination every day; determine the accessibility of your lab room and whether the experimental design

DATA SHEET FOR PART III: How Seed Germination Is Affected by Environmental Factors

Your question: _____

Your hypothesis: _____

At each time point, record the *number of germinated seeds* and calculate the *% germination* in each treatment.

Time elapsed	Description of treatments, and number of seeds in each treatment									
	# germ	% germ	# germ	% germ	# germ	% germ	# germ	% germ	# germ	% germ
t= 0 days										
t=										
t=										
t=										
t=										
t=										
t=										
t=										
t=										
t=										
t=										
t=										
t=										

would permit you to take the dishes home with you. Seeds kept in darkness should be checked quickly in dim light if the experiment is to continue. Record the treatments and observed data on the data sheet provided for Part III.

☐11. Graph your results. If you took data every 1–2 days, then produce line graphs of the *% germination* versus *time elapsed* for each treatment; plot these graphs on the same set of axes to facilitate comparison. To graph the final % germination for each treatment, use a bar graph.

☐12. Do a literature search on the effect of your environmental factor on seed germination. Prepare a written or oral report, as instructed, presenting and discussing your results and how they compare to the literature. (If several groups investigated a single variable with each group testing the same conditions of the variable, then be sure to compare your group's results to the pooled results at the end; which do you consider more reliable—your data or the pooled data? Why?)

FURTHER INVESTIGATIONS

Some sources claim that lettuce seeds need light to germinate. Other sources state that certain varieties seem to be able to germinate in the dark. More scientific sources give more precise information; for example, Bewley and Black (1994) stated that Grand Rapids lettuce seeds generally cannot germinate in darkness above about 23 degrees C, but can germinate in the dark below this temperature. They also state that lettuce seeds are affected by only brief exposure to white light—a few minutes or seconds. Saini et al. (1989) provided a more detailed analysis of the effect of temperature and various light conditions on Grand Rapids lettuce seed germination.

Design an experiment to clarify the factors that trigger germination in a particular lettuce variety: At what temperatures can the seeds germinate in the light? In the dark? If you find a temperature at which the seeds do not germinate in the dark, how much light exposure will trigger germination?

Perform the experiments if instructed to do so.

LITERATURE CITED

Bewley, J. D., and M. Black. 1994. *Seeds: Physiology of Development and Germination*, 2nd ed. New York: Plenum Press, 236–237.

Saini, H. S., E. D. Consolacion, P. K. Bassi, and M. S. Spencer. 1989. Control Processes in the Induction and Relief of Thermoinhibition of Lettuce Seed Germination: Actions of Phytochrome and Endogenous Ethylene. *Plant Physiology* 90:311–315.

7 PLANT DEVELOPMENT

LAB

I. Pet the plant: thigmomorphogenesis
II. Seedlings and light, revisited

BACKGROUND

A particular seed will develop into a mature plant of a genetically predetermined form—sunflower seeds always grow into sunflower plants. However, structural details such as exact plant height, stem thickness, stem orientation, and leaf size are influenced by various external factors (soil fertility, water availability, light condition, climatic conditions, and contact with other organisms) as well as various internal factors (hormones and other molecules).

External factors affect internal factors that affect the plant's growth and development. When a seedling emerges from the soil, light affects the form of the **phytochrome** (Greek: *phyton* = plant, *chroma* = color) molecules in the seedling, which in turn reduces the levels of the plant hormone ethylene produced by the seedling, which in turn straightens the growing tip and slows the stem growth. A plant growing in phosphorus-poor soil has an internal phosphorus deficiency, which is reflected in stunted growth; phosphorus is required in many cellular processes. The molecular details of plant developmental processes are very active areas of research in which much is still to be discovered.

PART I:
PET THE PLANT: THIGMOMORPHO-GENESIS (5–6 weeks)

OBJECTIVES

1. Practice the scientific method by properly designing and performing your own experiment.
2. Determine the effect of touch on the development of a plant species.
3. Gain experience in statistically analyzing data.

BACKGROUND

Consider these observations: An object touching a plant tendril on one side causes the tendril to start bending around the object within the hour. Plants growing between the rocks on a windy mountaintop are stunted compared to other individuals of the same species that grow in less-windy locations at lower elevations. A plant grown in the protection of a greenhouse tends to have longer, thinner stems than another plant of the same species that is grown in a garden where it is battered by storms and brushed against by animals and other plants.

These observations are all examples of plants responding to various types of touch. A tendril's bending in response to touch is called **thigmotropism** (Greek: *thigma* = touch; *trope* = a turning). **Thigmomorphogenesis** (Greek: *morphe* = form, *genesis* = development) is the development of form in response to touch. Thigmomorphogenesis in plants has been actively studied since the 1970s.

NOTE: Your instructor may have the class do only certain parts of this lab, or may have different student groups do different parts and then report their results to the class.

Now consider the following: If you are performing an experiment in which you track the growth of a plant over a period of time, *will your handling of the plant while taking frequent measurements actually change that plant's growth rate?*

The investigation in Part I allows you to design your own experiment to explore thigmomorphogenesis, and then gives you an opportunity to analyze data using statistical methods that are commonly used in various fields of study.

MATERIALS

- *Arabidopsis thaliana* seeds, wild type (Carolina HT-17-7600, Ward's 86 V 8216)
- if desired, also *Arabidopsis thaliana* seeds, Erecta tall variety (Ward's 86 V 8226)
- potting soil in tub; if dry, add water and stir with trowel to moisten before use
- trowels; pots or yogurt containers
- if needed, large nails and hammers for making drainage holes in the bottom of containers
- labels and felt-tip markers
- metric rulers

INVESTIGATIONS (5–6 weeks)

☐1. Working in groups, decide on a question to investigate. For example:
   - How does rubbing and flexing the stems twice a day affect the height of an *Arabidopsis* plant?
   - How does rubbing the leaves twice a day affect the height of an *Arabidopsis* plant?
   - Is there a difference in the response from rubbing and flexing stems as compared to rubbing leaves?
   - Is there a difference in the height (percent of normal untouched height) achieved by the rubbed and/or flexed wild-type as compared to rubbed and/or flexed tall-variety of *Arabidopsis*?

☐2. On Data Sheet 1, record your question and hypothesis.

☐3. Consider the following points when designing the details of your experiment:
   - Remember that the hallmarks of good experimental design are *replication*, a *control*, and a *single variable*.
   - Will all groups in class be doing the same experiment, with the data to be pooled at the end in order to have lots of replication? Alternatively, will two or more questions be investigated by the class, with several student groups per question in order to increase replication, with final oral presentations of results by the groups to the class?

- Is there an advantage to having each student taking control plants as well as experimental (to-be touched) plants home, rather than having the controls in the campus greenhouse and the experimental plants in students' homes?
- Decide on where in your room or home you could keep the plants. They need to be in the light, undisturbed, and readily available for your touch treatments. Each plant should not be touched otherwise—by other plants, curtains, or pets!
- How will you water the plants so that they are not touched by falling water drops and are receiving adequate moisture?
- Decide on the times for the twice-daily touch treatments—upon arising and at bedtime, perhaps?
- Decide on how and when to measure plant height such that you do not introduce additional handling that could influence the results. Will you just record final plant height after 5 or 6 weeks?

☐4. Present your group's experimental design to the class for discussion and suggestions. After the discussion, revise your experimental design.

☐5. Make sure that enough materials are on hand for all groups to do their experiments; if not, then either acquire the materials or modify your experimental design.

☐6. Set up your experiment. *NOTE*: Undersized seeds produce undersized seedlings; therefore, discard all undersized seeds and plant only normal seeds of the same size in the experiment.

☐7. After 5 or 6 weeks, measure the final height of the above-soil portion of each plant.

☐8. All students having done the same experiment will record all the raw data from that experiment onto the data sheet provided; then the total raw data can be analyzed.

☐9. Do the following basic data analysis:
   a. Calculate the mean height for the control plants and the mean height for the experimental (touched) plants (see Appendix 3, Part I); record the means in Table 7.1.
   b. Represent your raw data graphically: On one set of axes, plot a histogram of control plant heights as well as a histogram of experimental plant heights (see Appendix 2 for information on histograms).

DATA SHEET 1. Thigmomorphogenesis

Your question:

Your hypothesis:

Your experimental design:

7

Materials needed:

Final height of plants after ___ weeks		
	Heights of control (untouched) plants	Heights of experimental (touched) plants
Student 1		
Student 2		
Student 3		
Student 4		
Student 5		
Student 6		
Student 7		
Student 8		
Student 9		
Student 10		
Student 11		
Student 12		
Mean		
Standard Deviation		
Mean ± 1SD (see Appendix 3, Part I)		

Space for calculations:

Table 7.1 Raw data: thigmomorphogenesis

10. You may also be asked to do the following more advanced data analysis, or it may be offered for extra credit:

- In science, the mean value is generally presented in tables as "the mean, plus or minus one standard deviation." Calculate the standard deviation (see Appendix 3, Part I) for the control data and for the experimental data; record them at the bottom of Table 7.1. Also record the full expression "Mean ± 1SD" for the control data and for the experimental data.

- To determine whether there is a *statistically significant difference* between the means of the control and the experimental samples, you would have to do a two-tailed, two-sample t-test (see Appendix 3, Part II). This statistical test is commonly used to analyze data in many fields of study. This lab gives you the opportunity to practice using the t-test.

   The *two-sample t-test* is used when the two samples are totally independent of each other; the *paired-sample t-test* is used when each measurement of one sample is associated with (paired with) one and only one of the measurements of the other sample. If all plants were kept in a greenhouse, the two-sample t-test would be appropriate. If each student kept a control and an experimental plant on a windowsill at home, the paired-sample t-test would be more appropriate because each windowsill has different environmental conditions that could affect plant growth.

   On a separate sheet of paper, do your calculations and state your conclusion—whether there was a statistically significant difference between the untouched control plants and the touched experimental plants.

11. Your instructor may require an oral or written report, presenting and discussing your results as well as the implications for plant growth in nature.

Q1.   How would plants with a developmental response to touch have a selective advantage—a survival advantage in nature? Discuss at least one example specifically.

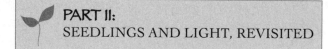

PART II:
SEEDLINGS AND LIGHT, REVISITED

II.A. HOW DO EMERGING SEEDLINGS PROTECT THEMSELVES?

OBJECTIVES

1. For various kinds of seeds, determine through observation which structure of the emerging seedling takes the wear and tear of pushing up through the rough soil.
2. Observe the changes that occur in the seedling's form when it emerges from the soil.

BACKGROUND

The plant embryo in a seed has one or more fleshy **cotyledon** structures attached to it (see Figure 7.1). For example, a peanut that you pop in your mouth is a seed; the peanut consists of two halves held together by a little nub; the two halves are the cotyledons that are attached to the little nub, which is the embryonic plant. In "split pea soup," the split peas are a pea seed's two cotyledons that have broken apart. See Lab 6 (Seed Germination) or Lab 18 (Reproduction of Flowering Plants) for more detailed information on seeds.

In some plant species, a seedling emerging from a seed has its embryonic leaves rolled up and encased in a protective sheath or **coleoptile** (Greek: *koleos* = sheath; *ptilon* = feather; like the sheath surrounding a young feather). Once the coleoptile has broken through the soil surface, its growth no longer keeps pace with the growth of the enclosed leaves, resulting in the leaf tips growing through the top of the coleoptile.

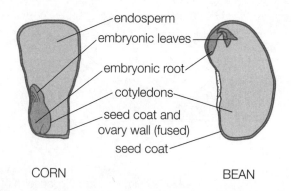

endosperm
embryonic leaves
embryonic root
cotyledons
seed coat and ovary wall (fused)
seed coat

CORN   BEAN

Figure 7.1 The parts of a seed

In other plant species, the **hypocotyl** (the seedling's axis that is *below* the point of attachment of the fleshy cotyledons; Greek: *hypo* = under, beneath) grows into a bend. This bend or **hypocotyl hook** is pushed up through the soil, dragging after it the cotyledons. The delicate embryonic leaves lie well protected between the large cotyledons.

In yet other species, it is the **epicotyl** (seedling axis *above* the point of attachment of the cotyledons; Greek: *epi* = on, over) that grows into a hook. The **epicotyl hook** pushes up through the soil, and shields the embryonic shoot tips which is pulled along behind.

In this investigation you will determine which method of protecting the embryonic leaves is used by emerging seedlings of peas, beans, and corn or wheat.

MATERIALS

• seeds of normal pea, bean, wheat or corn; or seeds of super-dwarf pea, soybean and wheat (Consult the website of the Crop Physiology Laboratory at Utah State University http://www.usu.edu/cpl/outreach_seed_info.htm for seeds of super-dwarf plants—Apogee wheat, Hoyt soybean, Earligreen pea, and others—all of which are ready for harvest within 2 months of planting; instructions for growing these super-dwarf crops in the classroom are also provided on the website.)
• pots or yogurt containers
• if needed, large nails and hammers for making drainage holes in the bottom of containers
• potting soil in tub; if dry, add water and stir with trowel to moisten before use
• trowels, labels, and felt-tip markers

INVESTIGATIONS

☐1. If there are no drainage holes in the bottoms of the containers, use a hammer and nail to make them. Add moist soil to 3 containers and gently tamp it down with your fingers; repeat until the soil level is 1–2 cm below the rim.
☐2. Plant a pot of each kind of seed, using 2 to 3 seeds per pot.
☐3. Take your pots home with you, and *keep the soil moist* (you could set the pots in a tray of water).
☐4. Closely check the pots for seedling emergence *at least twice a day* (each morning and each evening, at least).
☐5. When seedlings of each species begin to emerge, observe carefully and record your observations in Table 7.2.

□6. After observing the emerged seedlings, you can continue to observe their growth. If super-dwarf varieties were used, then you can observe the whole life cycle of these plants because they grow from seed to harvest within two months.

Q2.　Based on these observations alone, can you determine whether the changes in the seedling's form upon emergence from the soil were due to the absence of contact with soil, due to the presence of light, or due to some other factor?

II.B. EFFECT OF LIGHT ON SEEDLING DEVELOPMENT

BACKGROUND

Precollege science classes commonly observe emerging seedlings and their bending toward light (**phototropism**, from Greek: *photos* = light; *trope* = a turning). These experiments typically use colored plastic sheets. There are actually more factors involved in these experiments than meet the eye. Let's take a closer look.

	Pea	Bean	Wheat or Corn
Which part of the seedling broke through the soil surface?			
If a hook was present, did it persist or straighten after emerging from the soil?			
Did leaves enlarge greatly before or after emerging from the soil?			

Table 7.2 Observations of seedling emergence

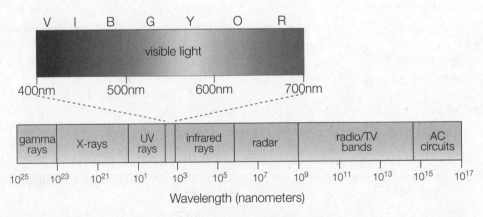

Figure 7.2 The electromagnetic spectrum. The letters VIBGYOR on the visible light spectrum indicate the location of violet, indigo, blue, green, yellow, orange, and red, respectively.

Visible light is a small portion of the large spectrum of radiation known as the electromagnetic spectrum (see Figure 7.2). All radiation has wavelike properties, with the distance from the peak of one wave to the peak of the next being called the **wavelength**, which is measured in nanometers (nm). Visible light ranges from violet (about 400-nm wavelength) through indigo, blue, green, yellow, orange, to red (about 700-nm wavelength).

Different colors of plastic sheets (filters) differ in the *type of wavelengths* transmitted. A blue filter looks blue to us because it does not absorb, but rather reflects and transmits a large amount of the blue wavelengths to the eye. In addition to the blue light, however, a blue filter may also transmit various amounts of some of the other portions of the light spectrum. Different shades of blue filter differ in light transmission properties. The Roscolux filters used in theater lighting come in many colors; the Rosco company website at http://www.rosco.com/us/filters/roscolux.asp provides a *spectral energy distribution curve* showing the wavelengths of color that are transmitted by each Roscolux filter.

You can use either of the following methods to determine whether a blue filter transmits purely blue light: (1) Hold the blue filter across the opening of a **spectroscope** or a **spectrometer**. Looking through the spectroscope or spectrometer, you can see the portions of the spectrum that are transmitted and thus visible; the portions of the spectrum that are blacked out (missing) have been absorbed by the filter. (2) Use a much more expensive machine, a **spectrophotometer**, to measure precisely the absorption of the filter at regular intervals across the range of 400- to 700-nm wavelengths, and then graph your filter's absorption spectrum across that range.

Filters differ not only in the type of wavelengths transmitted, but also in the *amount of light* transmitted. Thus it is possible that a seedling bends toward one color rather than another not because it was stimulated by that wavelength, but rather because that filter transmitted a greater amount of light. **Light meters** can measure the amount of incident light.

Yet another factor that can influence such experiments is the *type of light source* that is used—incandescent, cool white fluorescent, wide-spectrum plant growth lights, and so forth differ in the amount of each color of light that is emitted. For example, incandescent bulbs do not emit much blue light. The websites of Sylvania, General Electric and Philips provide the **spectral power distribution curves** of their various lamps.

Beyond this brief introduction, further details concerning light characteristics get rather technical. The websites of the lighting manufacturers provide further interesting technical background information if you wish it.

MATERIALS

- seeds of normal pea, bean, wheat, or corn; or seeds of super-dwarf pea, soybean, and wheat (Consult the website of the Crop Physiology Laboratory at Utah State University at http://www.usu.edu/cpl/outreach_seed_info.htm for seeds of super-dwarf plants—Apogee wheat, Hoyt soybean, Earligreen pea, and others—all of which are ready for harvest within two months of planting; instructions for growing these super-dwarf crops in the classroom are also provided on the website.)
- light meters (Carolina has several)
- spectroscope (Carolina HT-75-5319, Ward's 25 V 4996) or spectrometer (Carolina HT-61-2548, Ward's 25 V 5005)
- pots, yogurt containers, or 2-liter soft-drink bottles that are cut off at the shoulders
- if needed, large nails and hammers for making drainage holes in the bottom of containers
- potting soil in tub; if dry, add water and stir with trowel to moisten before use
- trowels, labels, and felt-tip markers
- foil, scissors, and black tape
- Roscolux filters: Italian blue #370, Alice blue #378, brilliant blue #69, moss green #89, fire #19, medium yellow #10 (A Roscolux-Supergel Swatchbook is also available from the company.)

INVESTIGATIONS

☐1. The class as a whole will investigate the following questions as well as any additional questions that the students wish:
- Are the changes that occur in the seedling's form upon emergence from soil due to the absence of contact with soil or due to exposure to light? How do seedlings grow and develop when they emerge from soil into total darkness as compared to light? What changes in growth and development occur upon exposure to light?
- Does any kind of light trigger these changes in a seedling's form upon emergence from the soil, or is it a certain wavelength (color) of light that is absorbed by a particular pigment, which in turn triggers these changes?
- If emerging seedlings are provided with light from the *side* rather than from above, will all seedlings

exhibit positive phototropism (bend toward the light)?

- Is it a certain wavelength (color) of light that triggers the positive phototropism?
- Do the different shades of blue produce the same result? Can any observed differences in seedling response be explained by the spectral power distribution curves of the filters used?

☐2. Work in groups, with each group selecting the question to investigate.

☐3. On Data Sheet 2, record your question and hypothesis.

☐4. When your group designs the details of your experiment, remember that the hallmarks of good experimental design are *replication*, a *control*, and a *single variable*. Decide on the type of data that will be recorded and when it will be recorded. Decide on the type and quantity of seeds used, how many per pot, where the pots will be located, and under what light source.

☐5. If you are working with colored light, decide on the design of the chamber to be used.

- Will the filter cover an opening in a light-tight box, or a foil-wrapped, 2-liter bottle that is cut off at its shoulders, or some other chamber? How big will the opening be?
- What will be the quantity of light that gets through the filter and to the soil surface; and how can you measure that with the light meter without admitting additional light into the chamber?

- Can you make the amount of incident light the same in each chamber (to eliminate light intensity as a variable) by altering the distance of the light source from the filter?

☐6. Present your experimental design to the class for discussion and suggestions. After the discussion, revise your experimental design.

☐7. Make sure that enough materials are on hand for all groups to do their experiments; if not, then either acquire the materials or modify your experimental design.

☐8. Set up your experiment.

☐9. Record your observations in Table 7.3. Look up the spectral energy distribution (SED) curve for your colored filters. Look at your light source, and determine its manufacturer and type of lamp; then look up its spectral power distribution (SPD) curve on the manufacturer's website or in other references.

☐10. In your textbook and other print references, read up on phytochrome, seed germination, and phototropism.

☐11. Your instructor may ask each group to make an oral presentation of the experimental results to the class. Include SED and SPD curves in the discussion of your results. Also include information learned from print references about the pigments and cellular processes that result in the observed changes in your seedling's development. Your instructor may also require a written lab report.

Treatment	Observed results

Table 7.3 Effect of light on seedling development

68

DATA SHEET 2. Effect of Light on Seedling Development

Your question:

Your hypothesis:

Your experimental design:

7

Materials needed:

EXTENDED ASSIGNMENTS

1. Light is absorbed by pigments, which then trigger the phototropism and other growth and developmental processes of a plant body that promote survival in the environment. Light also is absorbed by the chlorophyll and carotenoid pigments that are essential for photosynthesis—the process that uses light energy to produce sugars, providing the plant with chemical bond energy. As long as we are dealing with light in this lab, consult the chapter in your textbook that covers photosynthesis to find the figure showing the absorption spectra of chlorophylls and carotenoids. Look up the SPD curves of various kinds of lamps (incandescent lamps and cool white, daylight, and grow light fluorescent lamps at least) by consulting the websites of Sylvania, General Electric and Philips. What do the **grow lights** do that the others do not do on their own, and how does this benefit plant growth?

2. Go to the Plants-In-Motion website at http://plantsinmotion.bio.indiana.edu/plantmotion/starthere.html and look at some of the time-lapse videos of plant development; look under the headings photomorphogenesis, tropisms, nastic movements, and any others that you wish to explore. Read the narratives provided and watch the videos.

LITERATURE CITED

NOTE: If the web address has changed, do an Internet search for the company or organization producing that website, or search for the name of the website.

Rosco Laboratories. 2006. Rosco USA "Products, Filters, Roscolux," http://www.rosco.com/us/filters/roscolux.asp (accessed February 19, 2006).

General Electric Company. 2006. General Electric "Spectral Power Distribution Curves," http://www.gelighting.com/na/business_lighting/education_resources/learn_about_light/distribution_curves.htm (accessed February 19, 2006).

Hangarter, R. 2000. "Plants-In-Motion," http://plantsinmotion.bio.indiana.edu/plantmotion/starthere.html (accessed February 19, 2006). Bloomington: Department of Biology, Indiana University.

Koninklijke Philips Electronics. 2005. Philips "Philips Lighting," http://www.nam.lighting.philips.com/us/ (accessed February 19, 2006). On the Lighting page, perform a search for "spectral power distribution curve."

Osram Sylvania. 2000. Sylvania "Technical Information Bulletin: Spectral Power Distributions of Sylvania Fluorescent Lamps," http://ecom.mysylvania.com/miniapps/lightingcenter/PDFs/faq0041-0800.pdf (accessed February 19, 2006).

USU Crop Physiology Laboratory. 2005. "Outreach: Super-dwarf Seed Information," http://www.usu.edu/cpl/outreach_seed_info.htm (accessed February 19, 2006). Logan, UT: Crop Physiology Laboratory, Utah State University.

LAB 8

VEGETATIVE PROPAGATION AND PLANT CARE

I. Get plants for free!
II. Keep plants looking great!
III. How toxic is that pesticide?
IV. Alternatives?

PART I:
GET PLANTS FOR FREE!

OBJECTIVES

1. Learn when and how to perform various methods of plant propagation.
2. Actually perform several methods of plant propagation.

BACKGROUND:
VEGETATIVE PROPAGATION

Vegetative propagation, a type of cloning, consists of producing new independent plants from an original plant without going through sexual reproduction. Vegetative propagation commonly occurs in various plants in nature and serves as a quick way to colonize an open area. *Willow* twigs that are broken off and swept downstream to another muddy bank will develop roots and colonize the new site. *Multiflora rose*, an invasive alien from eastern Asia, can spread vegetatively when the end of a cane touches the ground and develops roots, anchoring the growing tip there. *Aspen* trees are called a **clonal** species because they commonly develop new sapling shoots vegetatively from horizontal roots of the original plant, resulting with time in a large grove of what appear to be individual trees. *Cholla* cacti produce stem segments that break off easily and fall to the ground, develop roots, and grow into new independent plants.

You can vegetatively propagate many plants by rooting sections of shoots or other plant parts. There are several *advantages* to vegetatively propagating your plants: (1) You get *many* plants *quickly*, without having

to wait for the plant to go through sexual reproduction in its life cycle to produce the next generation. (2) You can propagate *seedless* crop varieties such as bananas. (3) Your newly produced plants are *genetically identical* to the original plant. Commercial varieties of fruit trees, grapes, and so forth are vegetatively propagated to ensure that the new plants are genetically identical and thus their fruits have the characteristic flavor and appearance of that variety. (The fruit tissue that we eat is tissue from the mother plant that is bearing the fruit. The pollen-donor (male) parent contributes genes to neither the fruit tissue nor the seed coats, but only to the contents of the seeds within the fruit, affecting the traits of the seedling that will grow from these seeds.) Many commercial varieties of houseplants and landscaping plants are also vegetatively propagated to produce genetically identical plants.

Vegetatively propagated plants being genetically identical can also be a *disadvantage*. An orchard or grove of genetically identical trees can be decimated by a disease to which this tree variety is susceptible. In contrast, a genetically diverse population of trees may well include individuals that have some resistance to the disease.

In a plant, every cell has the same complete set of genetic information in its chromosomes; a particular cell in the plant uses a particular subset of its genetic information to perform its particular functions. Thin-walled cells called **parenchyma** (Greek: *para* = beside; *en* = in; *chein* = to pour) are cells that are "poured in beside" or, more correctly, occupy the space inside a plant around the vascular tissue and other structurally specialized cells. The parenchyma cells tend to retain the ability, under appropriate stimulation, to resume cell division and to differentiate; this is especially true of the parenchyma at wounds or at the **nodes** (nodes are the positions along the shoot where leaves are attached). Wounding at the nodes, such as ripping off

NOTE: Your instructor will indicate the parts of this lab that your class will perform.

the leaves, stimulates the parenchyma to differentiate into root initials. Application of one kind of plant hormone called **auxin** (Greek: *auxein* = to increase) further stimulates the initiation of these adventitious roots—roots that form on a plant where they normally would not appear.

Scientists are actively researching the process and control of root initiation. The rooting success of cut shoots (**shoot cuttings**) depends on the kind of plant. The cuttings of certain plant species are easily rooted, while those of other plant species are difficult or currently impossible to root.

Methods of Vegetative Propagation

Here are some general guidelines for vegetative propagation:

- Always keep soil moist while a cutting or plantlet is getting established.
- Loosely cover the cutting with plastic wrap for a few days to maintain humidity around the cutting. This reduces the water loss from the plant surface (called **transpiration**) and thus reduces wilting until new roots are formed.
- *Pinch off* with your fingernails the growing tip or apical bud in order to promote formation of side branches and thus a bushy plant.

Common methods of vegetative propagation are described and illustrated below:

Shoot cuttings (*Coleus*, Swedish ivy, *Zebrina*, etc.)

1. Cut off the shoot tip that includes several nodes; long shoots can be cut into several sections, each of which includes several nodes (nodes are places along the stem where leaves are attached).
2. On the cutting, remove the lowest leaf or two to increase the wounded area from which roots will develop.
3. Remove any flowers (energy for flowering should go into root formation instead).
4. Place the shoot into moist soil such that all wounded areas are beneath the soil.

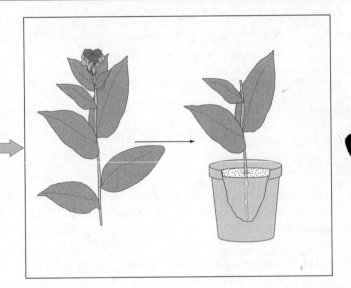

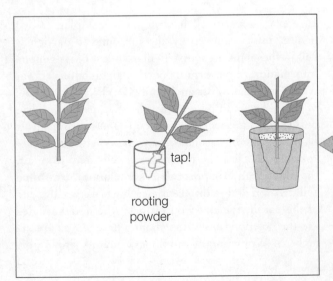

Shoot cuttings from a somewhat woody shoot (*Fuchsia*)

1. Cut off the shoot tip that includes several nodes.
2. Remove the lowest leaves.
3. Read warnings on the rooting powder package. *NOTE*: Do not inhale the powder.
4. Dip the cut end and wounded areas into the rooting powder (Hormodin®, Rootone®, or other brands).
5. Tap off all the excess powder! You need only a dusting of powder on the plant surface.
6. Poke a hole in the soil, insert the powdered base of the shoot, and firm the soil around it.

Stem cuttings (especially for thick-stemmed plants: *Dieffenbachia* and *Dracaena*)

1. Cut off a stem section having two or more nodes (places where a leaf was attached).
2. Lay the section on moist soil, or insert the *lower* end of the stem section vertically into moist soil.

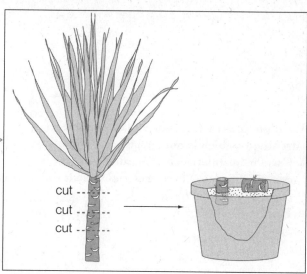

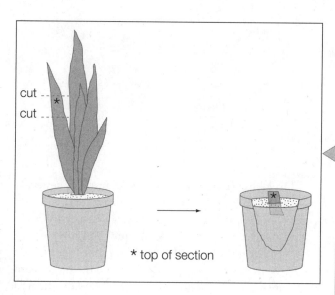

★ top of section

Leaf cuttings of *Sansevieria*

1. Cut the leaf crosswise to make 2- to 4-inch sections.
2. Place the *bottom* half of the section into moist soil. Polarity in the leaf prevents the top edge of the cut leaf section from developing roots.
3. Keep soil moist and be patient; it can take weeks for a shoot to form and emerge.

8

Leaf cuttings of fleshy-leafed plants (jade plant, African violet, *Peperomia*)

1. Cut a leaf from the plant.
2. Insert the leaf stalk into moist soil.

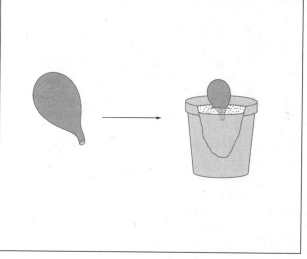

Offsets (side shoots of cactus, bromeliad, *Sansevieria*)

1. Cactus: Use tongs to gently break a side shoot/ pad off at a joint; place the base of this offset into soil.
2. Bromeliad or *Sansevieria*: Cut a side shoot off the parent plant and place it in moist soil.

73

Plantlets (*Kalanchoë*, spider plant, strawberry-geranium—which is not a true geranium)

1. Cut the plantlet from the runner of the spider plant or strawberry-geranium plant; break plantlets from the *Kalanchoë* leaf.
2. Put the plantlet's base with its root initials into moist soil.

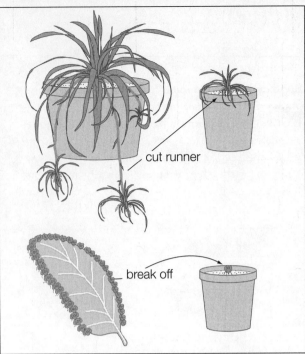

cut runner

break off

Divisions (plants growing in clumps of several shoots: prayer plant or African violet)

1. Remove the original plant from its pot by holding the pot upside down and tapping the pot edge on the edge of the table.
2. With your hands, gently separate the large plant into several smaller clumps and replant them in other pots.

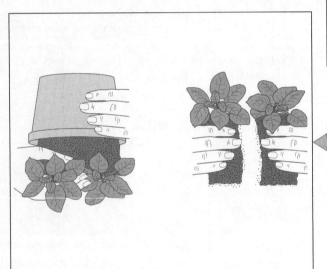

Air layering (large woody plants: fig tree, rubber plant)

1. Make cuts in one spot of the branch; dust it with rooting hormone powder.
   NOTE: Do not inhale the powder.
2. Using plastic wrap and a twist tie, form a loose, wide sleeve around the wounded spot and tie the sleeve tightly around the stem beneath the wounded spot.
3. Pack the sleeve with moist sphagnum moss; close the top of the sleeve around the stem using another twist tie.
4. Cut holes in the plastic to allow watering for the next several weeks; be patient and keep the sphagnum moist.
5. When roots are visible in the ball of sphagnum, cut the branch off and plant it.
6. The parent plant may be cut back drastically and kept watered; it will probably send out new side shoots.

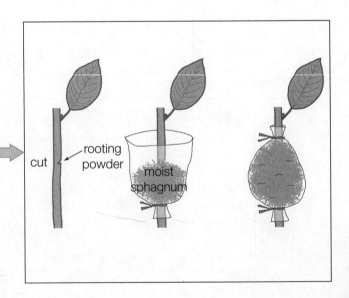

cut
rooting powder
moist sphagnum

74

MATERIALS

- containers: plastic pots or recycled plastic containers (yogurt cups, the bottom half of plastic soft-drink bottles, etc.)
- if needed, large nails and hammers for making drainage holes
- potting soil
- large tub in which to mix the dry potting soil with water until moist
- trowels
- gardening gloves, if desired
- rooting powder (Hormodin®, Rootone®, Hormex®, Stim-Root®, or other brands)
- plastic wrap
- plant labels or tape
- felt-tip markers
- tongs
- single-edge razor blades
- sphagnum moss
- twist ties
- a variety of plants, such as:
  - *Coleus*, Swedish ivy, *Zebrina*, or another soft-stemmed plant, indoors or outdoors
  - *Fuchsia* or another woody plant, indoors or outdoors
  - *Dieffenbachia* or *Dracaena*
  - *Sansevieria*
  - jade plant, African violet, or *Peperomia*
  - cactus or bromeliad
  - *Kalanchoë*, spider plant, or strawberry-geranium
  - prayer plant or African violet
  - fig tree or rubber plant

INVESTIGATION

Use at least three different methods to propagate at least three different plants.

□1. If using plastic recycled cups, make drainage holes using a hammer and nail.

□2. To each cup, add moist soil, tamp it down with your fingers, and repeat until the soil is about 1 cm below the cup's rim.

□3. Prepare your plant specimen as indicated in the background section.

□4. Insert the plant part to be rooted; firm the soil around it; water from beneath by placing the pot into a pan of water.

□5. Label the pot with the plant's name.

□6. Take your propagated plants home with you and keep the soil moist; you may also keep them loosely covered with plastic wrap for a few days to maintain humidity and reduce wilting until new roots are formed.

Q1. Why are cuttings or divisions often wilted in the first few days? (In your discussion, include answers to the following questions: How does a plant take up water? How does a plant lose moisture?)

8

PART II:
KEEP PLANTS LOOKING GREAT!

OBJECTIVES

1. Learn about selection and care of plants.
2. Learn to evaluate the condition of a plant's health and the suitability of its environment.
3. Learn to recognize some of the common insect pests.
4. Determine treatments that would improve and maintain a plant's health.

BACKGROUND: SELECTING AND CARING FOR YOUR PLANT

Have you ever visited someone whose house looked like a home for ailing and geriatric plants? This condition can be avoided by (1) initially selecting the correct plant for the location in the house, (2) properly caring for the plants, and (3) rejuvenating the plants as needed.

Select the Correct Plant

There has to be a good match between the environmental requirements of the plant species and the environmental conditions at its location in the house. (Plant species that are physiologically adapted to a shady environment will do rather poorly in bright sun, and vice versa.)

When selecting a new houseplant:
- Determine the *location in the house* for the new plant.
- Analyze the *light, humidity,* and *temperature* in that location.
- Analyze *yourself*: How often are you likely to remember to water your plants? If you want a plant that requires higher humidity than your location provides, are you willing and likely to frequently mist the plant with a sprayer and keep water in a tray of gravel on which the potted plant stands?
- Select the *right plant*—one that will grow well under those conditions of light, humidity, temperature, and frequency of watering. Consult houseplant books and pot labels for care requirements.
- *Quarantine* any new plant that you bring into the house; check the new plant for a few weeks for any insect pests that may hatch from tiny hidden eggs. If you cannot quarantine, then for several weeks carefully check the new plant and its neighboring plants for any new insect pests, and squash any pests that appear.

Care for Your Selected Plants

You can hover over your plants and tend to their needs daily, or you can provide the following basic care that will keep most plants in good shape:

- *Drainage holes* should be in the bottom of each pot so that excess water can drain into a tray underneath. Plants in decorative pots lacking drainage holes need to be monitored; roots in waterlogged soil gradually rot and the plant eventually dies.
- *Water your plant appropriately.* Consult a book on houseplant care to learn what your plant needs. *Peperomia* needs the soil to become moderately dry between waterings, while *Fittonia* needs the soil to be constantly moist.
- *Check for and squash any insect pests* while you water your plants; be sure to check the undersides of leaves and the tender growing tips—two favorite places for insects to hang out. Heavily infested areas can be cut off, bagged, and discarded in the trash outside—so that the pest cannot find its way back to the plant.
- *Fertilize* your plants occasionally; it makes a *big* difference! The "all-purpose plant food" works for most plants. There are also special formulations such as one for African violets, and another for acid-loving plants. Note that some plants need a rest period without fertilization in order to flower well thereafter. Consult a book on houseplant care for details on your plant.
- *Mineral deposits* eventually form as a crust on top of the soil of potted plants. Remove the mineralized upper layer of soil and replace it with fresh potting soil.

Rejuvenate Your Plants

You can give new vigor to your plants by pruning, propagating, or repotting them as needed.

- *Prune* (cut the shoot tips off) your gangly plant or sparse hedge to make it bushier, denser. By cutting back a shoot or "pinching" it off with your fingernails, you are removing the growing tip, which is the site of production of *auxin growth hormone.* Auxin from the shoot apex flows down the twig and *inhibits* the lateral (side) buds from growing into lateral shoots; this phenomenon is called **apical dominance**. If you remove the shoot tip, the lateral buds are released from inhibition and can start growing into shoots, making the plant bushier.
- *Propagate* your plant (as described in Part I of this lab) if all the lower leaves have dropped off your plant due to lack of fertilizer (various nutrients are

withdrawn from older leaves to nourish the growing tips) or if the plant is poorly shaped for some other reason. Root the cuttings from some leafy tips; discard the rest of the old plant, or cut it back severely and let it grow new shoots.

- *Repot* a pot-bound plant. A plant is pot-bound if roots are coming out of the drainage hole and the soil ball is full of and surrounded by roots. If root growth is limited, then aboveground growth is in turn limited. If you want to keep the plant growing well, then repot into a *bigger pot*. If you want to prevent the plant from getting much larger, then remove the plant from the pot (turn it upside down and tap the edge of the pot on the edge of the table), prune the roots by cutting a layer off the sides and bottom of the soil ball and repot with fresh soil around the root ball, in the *same pot*.

MATERIALS

- plants:
  - in the campus greenhouse or in your city's plant conservatory
  - in college buildings—lobbies, administrative offices, and so forth
  - at a discount store or nursery or garden center, especially plants on the "clearance sale" rack
- reference books on the identification and care of houseplants; there are many new ones in print, plus some very good older ones (available on the used book market) such as *Success with House Plants* (Reader's Digest 1979) and *The House Plant Expert* (Hessayon 1992).

INVESTIGATIONS

☐1. Tour the campus greenhouse or your city's conservatory with your instructor or the facility caretaker and learn to recognize the common insect pests: scale, mealybugs, whitefly, spider mites, and aphids.

☐2. Then, working in pairs, examine a variety of unhealthy-looking plants and evaluate their conditions (record your observations on the tally sheet provided):

a. What kind of plant is it, and what are its requirements? (Look it up in *at least two* of the references—does the information in various references agree?)

b. Is the plant (in campus buildings) getting too much or not enough light?

c. Is the plant in a temperature extreme (heat or air conditioner vent, or next to a cold window?)

d. Are there signs of inappropriate watering—wilted plant and hard, dry soil? Is the plant waterlogged? Is the pot draining properly?

e. Is the plant pot-bound?

f. Are there any insect pests present?

g. Are there mineral deposits on the soil?

h. Does the plant have a poor shape, requiring rejuvenation by propagation and/or pruning?

i. Your recommended treatment: Should the plant be discarded, or is it salvageable? What should be done to improve the plant's condition?

☐3. Present your evaluation of one of your plants to the class and describe the treatment you would recommend for your plant. During and after each presentation, the other students can examine the plant and see if they agree.

☐4. If you found conflicting information about a plant in the references, then do further research to try to clarify this. You could also interview your greenhouse manager or another horticulturist.

EXTENDED ASSIGNMENT (SERVICE LEARNING)

After finding and evaluating an unhealthy-looking plant in a campus building, offer to treat or rejuvenate the ailing plant and/or find a better location for it in the campus buildings.

8

TALLY SHEET FOR PLANT EVALUATION

Plant name:				
Environmental requirements of this plant:				
Light conditions present?				
Temperature conditions present?				
Signs of being over- or underwatered?				
Pot-bound?				
Insect pests present? Which ones?				
Mineral deposits on soil?				
Poorly shaped plant?				
Recommended treatment for this plant:				

PART III:

HOW TOXIC IS THAT PESTICIDE?
(information retrieval and discussion)

1. Learn about sources for information retrieval on pesticide toxicity.
2. Evaluate the sources of information.
3. Learn the meaning of *half-life* of a chemical.
4. Learn to think about the whole ecosystem and for the long term when considering the effects of chemicals.
5. Realize that there are still gaps in our knowledge about the effects of chemicals.
6. Realize that a chemical can have very different effects on various nontarget species.

BACKGROUND

What chemical are you spraying on your plants? Do insecticides affect only insects? Do herbicides affect only plants? What do you know about that chemical and its effects on you and other nontarget organisms in the ecosystem, over the short term and the long term? Where do you go for information?

In this lab you will first gather information about the toxicity of three pesticides that you, your family, or your neighbors use on plants; then you will evaluate your findings in a class discussion. But first, here is some terminology and background:

- **Pest**—any living organism that humans have decided is undesirable in a particular location. This can be any kind of organism—insects, plants, rodents, fungi, algae, and so on.

- **Pesticide**—literally means "pest killer." It refers to any chemical that kills or otherwise adversely affects a living pest organism. Pesticides, however, often affect more than just the target pest species in the environment. "All pesticides are toxic to some degree. . . . Never assume a pesticide is harmless" (EPA, 1992).

- **Half-life**—the length of time required for half of a chemical to be broken down. Note, however, that after one half-life, *half* the chemical is present, while after two half-lives, a *quarter* of the chemical is still present, and after three half-lives, an *eighth* of the chemical is still present, and so forth. Also note that the breakdown products may themselves be toxic.

- **LD50**—stands for 50% lethal dose; the amount of solid or liquid chemical administered in one dose that kills 50 percent of the test organisms; LD50 is usually expressed as mg/kg of a particular test organism indicated (mg of chemical per kg of body weight of the test organism). Note that the surviving 50 percent of organisms could be quite ill.

- **LC50**—stands for 50% lethal concentration; the amount of dust, gas, mist or vapor that kills 50 percent of the test organisms; LC50 can be expressed as ppm (parts per million in air), or mg/m^3 (milligrams per cubic meter of air; a milligram is 0.001 gram or 10^{-3} gram), or $\mu g/L$ (micrograms per liter of air; a microgram is 0.000001 gram or 10^{-6} gram).

- **"EPA Registered"**—when you see this statement on a pesticide package, it does not indicate that the chemical is harmless. The U.S. Environmental Protection Agency (EPA) awards "EPA registration" to a chemical after a risk assessment has been performed, evaluating the chemical *when used according to instructions* on particular crops or sites; the risk assessment weighs the benefits versus the risks, based on information available at that time. Controversies arise because different people have different concepts of "acceptable risk" as well as of the sufficiency of available data. A registered pesticide may become *restricted* or even *banned* later on due to the accumulation of new data documenting harmful effects.

According to the Federal Insecticide, Fungicide, and Rodenticide Act (FIFRA), all pesticides sold or distributed in the United States must be EPA registered; however, unregistered pesticides can still be legally used if an emergency exemption, an experimental use permit, or a state-specific registration has been granted.

Read more about this at the EPA website; there is a wide range of subheadings to click on, such as the Pesticide Registration Program (http://www.epa.gov/pesticides/factsheets/registration.htm), or you can go to the A–Z Index and click on "Banned pesticides" or other headings to learn more.

8

III.A. LOCATING INFORMATION (pre-lab homework)

NOTE: It is not unusual for web addresses to change over time; if one of the listed web addresses does not work, do an Internet search for the organization producing that site.

☐1. *Find three pesticides* that you, your family, a neighbor, or someone you know uses on the lawn, garden, trees, or houseplants. For example, look for the following:
- herbicides (weed killers); lawn treatments
- insecticides (against grubs, caterpillars and beetles, ants, hornets, tent caterpillars, etc.)
- fungicides (on seeds or applied to lawns or other plants)
- rodenticides

For each pesticide, use a separate Pesticide Worksheet provided in this lab; write down the *name of the pesticide*, its *manufacturer*, and its *active ingredient(s)*.

☐2. Consult the following databases and any other *reliable* website that you find. Summarize your findings on the worksheets provided in this lab; try to find each chemical in *more than one database*, so that you can compare information. Start with EX-TOXNET, since it is very readable:

a. http://extoxnet.orst.edu/ is the site of the EXtension TOXicology NETwork, which is a cooperative effort of several universities.
- Check out the EXTOXNET Frequently Asked Questions, to learn some background information.
- Click on "Search and browse" and conduct a search for your chemical name. *NOTE*: Not all chemicals in use today are in this database.
- Read the PIP (Pesticide Information Profile) for each of your chemicals. Note the date of last update.

b. http://www.pesticideinfo.org/Index.html is the site of the Pesticide Action Network North America. This site includes an extensive database of pesticides that you can search. Note the date of last update.

c. http://npic.orst.edu/ is the site of the National Pesticide Information Center (NPIC). Click on "Technical info" to find lists of websites for fact sheets and databases on chemicals and their toxicity.

☐3. What if you get a job working with this chemical routinely? Read the label fully, and also read the Material Safety Data Sheet (MSDS) for your chemical. The MSDS is meant not for consumers, but for workers exposed to this chemical in an occupational setting (i.e., higher concentrations, longer term). You can go to the manufacturer's website and often find their MSDS online, or get to your MSDS through links on the NPIC website or through the "MSDSonline" website (http://www.ilpi.com/msds/index.html), which also has links to tutorials on reading an MSDS.

Product name: _____

Manufacturer: _____

Name of the *active ingredient(s)*: _____

Website address(es) you are using as your information source on this active ingredient:

Acute toxicity:

Chronic toxicity:

Effects on birds:

Effects on aquatic organisms:

Effects on humans/mammals:

Environmental fate—the half-life and breakdown in different locations (soil, water, etc.):

An additional piece of information that you found interesting, such as the effect on bees, the mode of action, or the regulatory status:

PESTICIDE 2 WORKSHEET

Product name: _____

Manufacturer: _____

Name of the *active ingredient(s)*: _____

Website address(es) you are using as your information source on this active ingredient:

Acute toxicity:

Chronic toxicity:

Effects on birds:

Effects on aquatic organisms:

Effects on humans/mammals:

Environmental fate—the half-life and breakdown in different locations (soil, water, etc.):

An additional piece of information that you found interesting, such as the effect on bees, the mode of action, or the regulatory status:

Product name: _____

Manufacturer: _____

Name of the *active ingredient(s)*: _____

Website address(es) you are using as your information source on this active ingredient:

Acute toxicity:

Chronic toxicity:

Effects on birds:

Effects on aquatic organisms:

Effects on humans/mammals:

Environmental fate—the half-life and breakdown in different locations (soil, water, etc.):

An additional piece of information that you found interesting, such as the effect on bees, the mode of action, or the regulatory status:

III.B. EVALUATING YOUR FINDINGS (class discussion)

1. Did anyone find conflicting information on different websites for the same chemical? Why might this be? When was each site last updated? What scientific research studies were cited, and from what years? Who produced each website, and who funds that group? For each website that you consulted, do you feel that it is totally unbiased? What are the pros and cons of using a website versus published material?

2. Did you find any chemicals that had very different effects on different kinds of animals (different effects on fish than on birds, for example)? Tell about these findings, and explain why these differences might exist.

3. Why is the toxicological effect on bees frequently studied? What are the ramifications of killing the bees?

4. What was the shortest and what was the longest half-life that the class found? Did the half-life differ in different environments for the same chemical? Why might this be?

5. Did you come across comments such as "No data are currently available" for some toxicity categories? For the categories where there were data, how well studied were those toxicological categories? Was there only a single study? How adequate do you feel that one study is, and why?

6. How much information did you find under "chronic toxicity" as compared to "acute toxicity"? Why? How transferable do you think the findings in an animal study are to humans? Did you find any chemical that was toxic to one species but rather benign to another species of the *same* animal group (different effects on different species of fishes, for example)?

7. What terminology or concepts did you not understand on the websites? Can the class, with guidance of the instructor, figure those out?

8. Did anyone come across any chemical that was totally nontoxic to all nontarget species, and toxic to only one pest species? What are the ramifications of toxicity to nontarget species in the ecosystem and to the predator-prey interactions in the ecosystem?

9. If a chemical has a half-life of six months, how much of the chemical is left in the environment after two years?

EXTENDED ASSIGNMENT

Plant biology is connected to crop production, which is connected to pesticide use and worker safety. Become more globally informed by doing an Internet search for the keywords *pesticides and farm workers*, as well as other keywords. You will find that the Food and Agriculture Organization (FAO) of the United Nations is very concerned about the pesticide poisonings and deaths of farm workers in the developing world. Pesticide poisoning of farm workers occurs at a much lower rate in developed countries. Find out why.

PART IV:
ALTERNATIVES?
(homework and discussion)

OBJECTIVES

1. Learn about less chemically intensive methods for dealing with a plant problem.
2. Gain a better understanding of people's attitudes toward weeds.

IV.A. ALTERNATIVES TO USING INSECTICIDES ON HOUSEPLANTS

Determine the situations in which you, your family or your neighbors have used pesticides on plants in the past. Could the problem have been solved by a method other than pesticides? (Note that certain kinds of houseplants are adversely affected by certain insecticides; the plants look rather "sick" after having been sprayed.)

Source of a Pest Outbreak

When your houseplants have problems with insect pests, ask yourself these questions to determine the factors that permitted the outbreak:

1. Were the insects introduced to your plants due to a newly purchased plant not having been quarantined for several weeks?
2. Could the insect problem on your houseplant have been "nipped in the bud" had you more closely monitored your plants and squashed the first insects when you saw them?
3. Could you have pruned back and discarded the infested growing tips of the plant, and then closely monitored the rest of the plant and squashed any insects that appeared?

Alternatives to Insecticides for Dealing with a Pest Outbreak

Many companies now sell **beneficials**—beneficial insects that either feed on or parasitize particular pest species. The pest insect species must first be correctly identified, however, so that you can select the appropriate beneficial insect species.

Some old remedies have persisted because they also successfully control insects. For example, to successfully control scale, mites, and mealybugs: (1) Make a slimy solution of pure soap flakes (not detergent) in water; spray it onto the plant, covering the plant well. (2) Within one hour, thoroughly hose off the solution with a fairly intense spray of water. This old but effective method was contributed by Phil Davis from Cornell University on October 7, 2003, to the AERGC Forum. The Forum is the e-mail discussion group of the AERGC (Association of Education and Research Greenhouse Curators; home page is at http://www.life.uiuc.edu/aergc/); subscription to the Forum is free, as is the searching of AERGC archives by subscribers.

Other effective products, such as worm castings, are also on the market. That rather interesting topic has also been discussed in the AERGC Forum.

INVESTIGATION

☐1. Select an insect pest on plants, such as various kinds of scales, mealybugs, mites, aphids, whitefly, thrips, and so forth. You can consult a book on plant care to see the diversity of insect pests on houseplants.

☐2. Find information on your chosen insect—how does it damage a houseplant, what is its feeding method, and are there any control methods of low or no toxicity? Some keywords would be *biocontrol*, *beneficials*, *beneficial insects*, and *IPM (integrated pest management)*.

☐3. Present your findings in written or oral form, as instructed. Evaluate the effectiveness, toxicity, cost, and sources of materials for the methods.

IV.B. ALTERNATIVES TO HERBICIDES ON LAWNS

What do you view as an "aesthetically acceptable" lawn? Why? Does it have to be a monoculture of one kind of grass only? Why?

Dandelion: Why is it the most hated of all lawn "weeds" in suburbia?

The term **weed** is a human designation; a "weed" is simply a plant that is growing where a human does not want it to be growing. Weeds usually grow very well without any assistance from humans.

Why do so many people hate dandelions? Consider the following: Do dandelions have an ugly flower? Does their flower stink? Do they have unattractive fruits? Do they give you a rash when touched? Do they have spines or thorns on the plant or on the seeds that can get into bare feet? Are they poisonous?

8

The dandelion root, flower buds, flowers, and vitamin-rich leaves are edible (see Peterson, 1977), and the flowers yield an excellent honey. These plants have been grown commercially for their leaves and their roots. The dandelion also has a long history of being considered a medicinal plant.

You may have noticed that dandelions go through a mass flowering event around May of each year, producing a sea of yellow blossoms in the lawns. During the rest of the summer, however, only occasional blossoms are produced by the dandelion plants; therefore, throughout the summer, an unsprayed lawn and a lawn sprayed with herbicides look pretty much the same from a distance—solid green.

READ

In Robert Fulghum's 2003 book of brief essays entitled *All I Really Need to Know I Learned in Kindergarten*, read the essay entitled "Dandelions"; it's well worth it! On page 200, Fulghum pretty much sums up people's attitude about plants:

> "If dandelions were rare and fragile, people would knock themselves out to pay $24.95 a plant, raise them by hand in greenhouses, and form dandelion societies and all that."

DISCUSS

1. What forces have shaped people's perceptions, leading them to hate a harmless plant—and hating it to the point of applying herbicides to lawns every year? What forces have shaped people's perceptions so that they prefer a uniform green lawn to a lawn speckled with various wildflowers? If you have a negative attitude toward dandelions, at what age did it start, and why?
2. Discuss the power of advertising: What is the effect on people of being repeatedly exposed to ads about getting rid of "unsightly" clover or about dandelions being the enemy? What is the effect of peer pressure among suburban homeowners? How strong is peer pressure in your neighborhood? On your campus?

LITERATURE CITED

Association of Education and Research Greenhouse Curators (AERGC). 2004. "AERGC Forum," http://www.life.uiuc.edu/aergc/ (accessed February 19, 2006).

Extension Toxicology Network (EXTOXNET). 2006. http://extoxnet.orst.edu (accessed February 19, 2006). Corvallis, OR: Oregon State University.

Food and Agriculture Organization of the United Nations (FAO). 2006. "Newsroom," http://www.fao.org/newsroom/en/news/2004/index.html (accessed February 19, 2006). Rome, Italy: FAO.

Fulghum, R. 2003. *All I Really Need to Know I Learned in Kindergarten: Reconsidered, Revised & Expanded with Twenty-Five New Essays*, 15th anniv. ed. New York: Ballantine ("Dandelions," pp. 199–201).

Hessayon, D. G. 1992. *The House Plant Expert*. New York: Sterling Publishing.

Interactive Learning Paradigms, Incorporated (ILPI). 2006. "MSDSonline," http://www.ilpi.com/msds/index.html (accessed February 19, 2006). Lexington, KY: ILPI.

National Pesticide Information Center (NPIC). 2005. http://npic.orst.edu/ (accessed February 19, 2006). Corvallis, OR: Oregon State University and U.S. Environmental Protection Agency (EPA).

Pesticide Action Network (PAN). 2005. "PAN Pesticides Database", http://www.pesticideinfo.org/Index.html (accessed February 19, 2006). San Francisco, CA: Pesticide Action Network North America.

Peterson, L. A. 1977. *A Field Guide to Edible Wild Plants of Eastern and Central North America* (Peterson Field Guide Series). Boston, MA: Houghton Mifflin.

Reader's Digest Editors. 1979. *Success with House Plants*. Pleasantville, NY: Reader's Digest.

U. S. Environmental Protection Agency (EPA) Office of Prevention, Pesticides, and Toxic Substances. 1992. Healthy Lawn, Healthy Environment: Caring for Your Lawn in an Environmentally Friendly Way. Publication number 700-K-92-005. Washington, DC: U.S. EPA. Also available online at http://www.epa.gov/oppfead1/Publications/lawncare.pdf (accessed February 19, 2006).

U.S. Environmental Protection Agency (EPA). 2004. "Pesticide Registration Program," http://www.epa.gov/pesticides/factsheets/registration.htm (accessed February 19, 2006). Washington, DC: EPA.

TREE INVESTIGATIONS: Broad-Leaved Trees

LAB 9

I. What do twigs tell you?
II. Have you read a good tree stump lately?
III. Forensics: what can a small stem fragment reveal?

PART I:
WHAT DO TWIGS TELL YOU?
(field trip)

Why bother knowing about the markings and structures on twigs?

- It enables you to identify trees in winter (in February, identify sugar maples to tap in order to make maple syrup).
- It enables you to evaluate a tree's health by seeing how much the twigs grew in each of the recent years.
- It enables you to predict how many flowers will be on your rhododendron, dogwood, or magnolia in the spring.

OBJECTIVES

1. Understand the contents of different kinds of buds (Part I.A).
2. Understand the source of the markings and structures on a twig and what they indicate (Part I.B).
3. Understand how a twig grows such that you can determine the amount of growth produced on a twig during each of several recent years (Parts I.B and I.C).
4. Practice using the t-test to evaluate data (Part I.C).

I.A. WHAT IS A BUD?

BACKGROUND

The **terminal** or **apical bud** is at the end of the twig; its growth will lengthen the twig. **Lateral** or **axillary buds** are along the side of the twig; their growth produces side branches that make the plant look bushier. The lateral buds are also called axillary buds because they form in the **axil** (angle) between the leaf and the twig; on winter twigs, the axillary bud is thus positioned just above (distal to) the leaf scar.

- *Big buds, little buds.* A **vegetative bud** contains a short, immature shoot with tiny immature leaves; a **flower bud** contains tiny immature flowers. On some plant species such as rhododendron and flowering dogwood, the vegetative buds and the flower buds are distinctively different in size and appearance. Other plant species do not have separate flower buds, but rather have **mixed buds** containing the immature flowers along with immature leaves. The terminal buds on the short lateral shoots (spurs) of apple trees are mixed buds.

- *"It must be spring; the trees are getting buds."* Actually, those buds have been there ever since last summer, when they were small and in the axil of every leaf and at the tip of the twig. In the springtime, the buds simply get larger and more noticeable due to the growth of the immature shoot inside each bud.

MATERIALS

- computers with Internet access
- landscaping plants on campus that produce flower buds distinguishable from their vegetative buds: doublefile viburnum, magnolia, rhododendron, mountain laurel, flowering dogwood, nannyberry, blueberry, peach tree, or other species indicated by your instructor

INVESTIGATIONS

☐1. Look at the online interactive version of the McIntosh Apple Development poster on the Botanical Society of America's website (http://mcintosh. botany.org/bsa/misc/mcintosh/mcintosh.html); note that if the website address has changed, do an Inter-

NOTE: This lab may cover more than one lab meeting, or your instructor may assign different parts to different student groups. The purpose of this lab is to provide an understanding of how a tree functions.

net search for the poster title and/or the society. The poster shows 20 photos of the progression from winter bud to mature apples and leaves. Click to view a close-up and descriptive paragraph of each photo.

☐ 2. In late summer through early spring, locate any of the listed trees or shrubs on campus and note their larger and/or plumper flower buds versus their smaller and/or thinner vegetative buds (if you are uncertain, consult images in a reference—a book or an Internet site).

a. From the abundance and distribution of flower buds on the plant, deduce how showy a display there will be in the next flowering season.

b. Pick a bud from an inconspicuous location on the plant and dissect the bud by using your fingernails and pen point outdoors, or by using dissecting probes and a dissecting microscope indoors. Does the bud include flowers or only leaves?

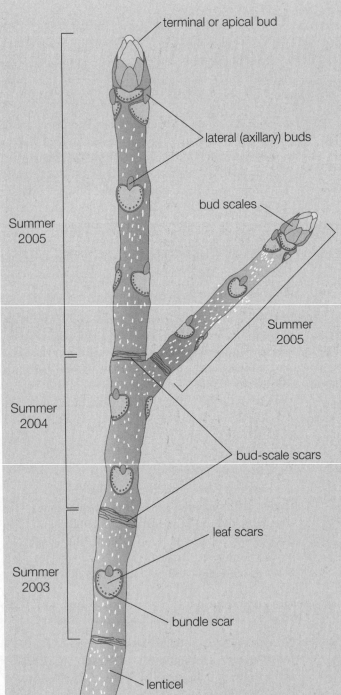

Terminal or apical bud
- Contains miniature unelongated shoot
- The twig grows in length when its terminal bud grows

Lateral or axillary bud
- Grows into a side shoot (lateral shoot)
- Present in every leaf axil, and above every leaf scar

Bud scales
- Tough, shingle-like modified leaves covering a bud
- Protect the bud contents against damage and desiccation
  - Oak bud has many bud scales
  - Birch bud has several bud scales
  - Willow bud has one cap-like bud scale
  - Pawpaw bud is a miniature hairy shoot with no bud scales

Bud scale scars
- Thin scars left behind when bud scales fall off in the spring

One year's growth
- The length of twig from a terminal bud back to the first set of bud scale scars
- The length of twig between two sets of bud scale scars

Leaf scar
- Scar left behind when a leaf falls off the twig
  - Maple twig has narrow, V-shaped leaf scars
  - Buckeye twig has large, shield-shaped leaf scars

Bundle scar
- Within a leaf scar
- The end of a vein (vascular bundle) that had entered the leaf

Lenticel
- Wart-like break in the twig surface
- Allows gas exchange
- Is linear, across the twig in birch and cherry
- Is a round dot in buckeye and maple

Figure 9.1 Anatomy of a twig, collected in autumn of 2005

1.B. TWIG CLUES

BACKGROUND

Study Figures 9.1 through 9.3 and refer to them during the investigation.

Telltale Clues!

Heed these clues:
- On a winter twig, the *leaf* arrangement can be figured out by looking at the *leaf scar arrangement*.
- You can deduce the *leaf arrangement* even if there are no low-hanging branches and it is winter, by looking at the lateral twig arrangement silhouetted against the sky. Opposite leaves have opposite buds that grow into opposite lateral twigs—unless a bud died or was eaten by wildlife.
- In the northeastern United States, the *native trees* with opposite leaves are the maple, ash, dogwood, and the horse chestnut/buckeye group; the memory aid for opposite-leaved *trees* is thus **M-A-D Horse**.

Things to Watch Out For!

Don't get tripped up by these features:
- **False terminal bud:** Some plants such as buttonbush, tree of heaven, and basswood have *no true terminal bud*. Toward the end of the growing season the apical growth aborts, and the bud that you see somewhat off center at the end is actually a lateral bud. Right beside this false terminal bud, you can see the stump of the twig's end.
- **Spur shoots:** Certain species such as the birches can produce some very short, slow-growing side branches called "spur shoots"; these are difficult to read because of the very closely spaced sets of bud scale scars. Select a normal terminal shoot rather than a spur shoot to examine in class.
- **No bud scale scars:** Some species like pawpaw trees have no bud scales and thus no bud scale scars. The bud consists of simply a short, immature shoot covered with hairs.
- **Hidden lateral buds:**
  - In sycamores, the bulging base of the leaf stalk completely surrounds and encloses the lateral bud. Break the leaf off carefully, and you will find the cone-shaped lateral bud.
  - In Kentucky coffee-tree, the small buds are sunken into the bark and are thus not obvious.

MATERIALS

- rulers (metric)
- trees outside

Figure 9.2
Leaf arrangements

Alternate leaves   Opposite leaves   Whorled leaves

Figure 9.3
Lateral buds and leaf scars help in
identifying trees

Oak   Maple   Willow   Ash

9

INVESTIGATION

☐ 1. Work in pairs. Locate and look at four different tree species.

☐ 2. Find and describe the various twig markings in the following chart for twig anatomy; consult the background information for help.

Q1. Did most of the twigs on most of the trees examined by the class show poorest growth in the same year? _____ If so, was there any abnormal environmental condition that year?

Q2. Why might one twig be different in growth pattern from most of the twigs on that tree?

Q3. A particular summer had good growing conditions, but a tree's twigs showed poor growth for that summer. What factors might have caused the poor growth?

CHART FOR TWIG ANATOMY

Tree Species				
Leaf scars opposite or alternate?				
Number of bud scales on a bud? (many, several, two, one, none)				
Shape of leaf scar; number and arrangement of bundle scars				
Lenticels—shape and abundance on twig				
Bud scale scars—many in a wide band, or few and hard to see, or none?				
Presence of true terminal bud?				
On your twig, which of the last 3 years produced the most (longest) twig growth?				
Look at 10 twigs; in the last 3 years of growth, was there one year that usually had the worst growth?				

1.C. TWIG GROWTH

In Part I.B, you probably noticed that each twig grew a little differently and that it can be difficult to determine whether there was a real difference overall in growth during one year as compared to another year. In science, when you want to determine whether there is a significant difference between two groups, you sample each group and then perform a statistical test on the data to determine whether there is a significant difference. The following investigation will give you practice in doing just that.

The purpose of this investigation is to determine whether a particular tree produced the same amount of growth in each of the last two complete growing seasons, or whether there was a significant difference in growth between those two years.

MATERIALS

- rulers (metric)
- trees outside

INVESTIGATIONS

☐ 1. The class will select a tree with easy-to-reach and easy-to-read twigs.

☐ 2. Each student will measure one or two twigs (depending on class size) to determine the length of twig growth in each of the past two full growing seasons. Record your data.

☐ 3. Compile the class data in the following class data sheet.

☐ 4. Each student will group the lengths into *size categories* (categories must be of equal size) and draw a bar graph (histogram) of the data for each year. Put both histograms superimposed on the same set of axes to facilitate visual comparison of the data (see Appendix 2, "Data Presentation"). Your instructor will tell you whether to do this by hand or whether the class will go to the computer lab to use a graphing software program such as Microsoft® Excel.

☐ 5. Determine whether there is *a significant difference* in growth between the two years. To determine this correctly, you need to do a *paired-sample t-test*, because each measurement of one year is associated with (paired with, taken from the same twig as) one and only one of the measurements of the other year. You may have noticed that on a branch, the leader twig tends to grow more than the lateral twigs. See Appendix 3 for instructions on the t-test, performed by hand and by Microsoft® Excel®.

CLASS DATA SHEET: Twig Growth

Twig Number	Length of Twig (cm) grown in year: 2007	Length of Twig (cm) grown in year: 2006
1	9.5	21.0
2	6.5	11.5
3	3.0	6.0
4	1.6	2.5
5	20.6	9.5
6	8.0	14.0
7	7.9	10.7
8	10.2	12.6
9		
10		
11	WEDNESDAY	
12		
13	16.5	15.0
14	26.5	31.75
15	3.9	7.2
16	14.5	13.0
17	6.5	9.0
18	13.0	16.0
19	5.1	5.9
20	3.5	3.1
21	15	19
22		
23		
24		
25		
26		
27		
28		
29		
30		
Mean (average) length		

☐ 6. If you find a significant difference in growth between the past two years, then research the environmental conditions during those two growing seasons and propose a factor that might have caused this difference in growth.

☐ 7. Summarize your findings in a report.

91

OBJECTIVES

1. Understand how a tree stem grows and functions.
2. Learn what the markings on a tree stump indicate and how to read the age and growth rates over the years.
3. Understand why hollow trees can still live.
4. Understand how a tree heals its scars.
5. Understand why a tree doesn't dry out even though the bark keeps ripping open.
6. Understand why the knots don't fall out of knotty furniture, but knots on some boards do fall out.

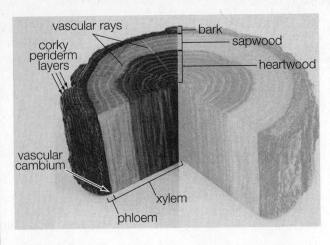

Figure 9.4 Section of a tree. The wood is divided into the heartwood and sapwood regions.

BACKGROUND

How Does a Tree Grow?

Plants do not grow in the same manner as animals do:

- Animals—grow until adulthood by cell division throughout their bodies. (Fish grow in this manner throughout their lives.)
- Plants—grow throughout their lives, but only in special localized regions called **meristems** (Greek: *meristos* = divided, divisible).
  - Growth in **length** occurs *only* at the *tips* of shoots and roots (the apical meristems).
  - Growth in **width** of that newly produced shoot or root occurs in all subsequent growing seasons.

Growth in Width

A tree grows in width in a precise manner:

- Growth in width occurs *only* in a thin layer of *lateral meristem* (the **vascular cambium**) located between the wood and the bark of shoots, branches, stems, and roots (see Figures 9.4 and 9.5).
- The vascular cambium produces the following two tissues:
  - New **phloem** tissue, added outwardly to make more inner bark (for transporting sugars from the leaves to wherever they are needed or stored).
  - New **xylem** tissue, added inwardly to make more wood (for conducting water and minerals up the plant).
- Thus, the most recently produced xylem is the outermost xylem, adjacent to the vascular cambium; the most recently produced phloem is the innermost phloem of the inner bark.

- Another lateral meristem, the cork cambium, contributes only slightly to the growth in width and will be discussed later in this lab.

Q4. If you nail a trail marker at eye level onto a tree, will it slowly move higher as the tree grows taller? Will you have to come back in 10 years and nail it back at eye level? Will you need to lengthen the ropes of a child's swing each year to keep the seat the same height off the ground? Explain.

Growth Rings in the Wood (xylem)

Growth rings tell about the environmental conditions (refer to Figure 9.5):

- In early *spring* when growing buds need much water, large-diameter xylem cells are produced (early wood).
- *Late* in the growing season, smaller diameter cells are produced in the growth ring (late wood).
- **Growth rings** are visible in the wood because of the contrast between last year's small-diameter, dense late-summer growth, and this year's larger-diameter springtime growth, which is right next to it.
- Thus, you can *count the growth rings* and deduce the age of the tree. (Note, however, that certain tree species may have two growth spurts in a summer, producing a double ring.)
- The *thickness of a growth ring* reflects the growing conditions that year. Drought, insect attack, disease, too much shade result in thin growth rings.

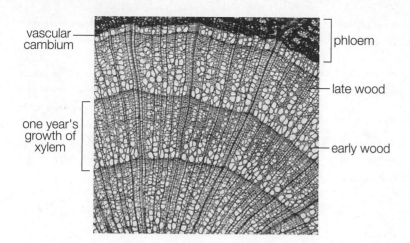

vascular cambium

one year's growth of xylem

phloem

late wood

early wood

Figure 9.5
Growth rings in cross-section, close-up

- Read a tree's record of growth rings from the outside inward, and go back into time deducing the growing conditions for each year for that tree.

Lopsided Rings

A *leaning tree* produces a whole series of consistently lopsided rings, all of which are wider on the same side of the tree. A leaning conifer tree produces more wood on the underside, while a leaning hardwood tree produces more wood on the top side.

Tree Rings Help Solve Historical Mysteries!

The tree rings of *living* trees can be studied by taking a narrow core of the tree by means of an instrument called an **increment borer**. The core resembles a thin dowel rod, across which the tree rings are arranged.

To learn about climatic conditions in an area throughout history, you would examine several ancient trees in that area because they will have experienced the same climate. Climatic effects would be reflected in the tree ring pattern held in common by these trees. An individual tree may show occasional deviations from the typical pattern for the area due to occasional nonclimatic factors.

One historical mystery is the "lost colony" established in 1587 on **Roanoke Island**; subsequent colonists found no trace of the original colony. Recent tree ring analysis of ancient bald cypress trees in the southeastern United States (Stahle et al. 1998) has indicated that the most extreme drought of the past 800 years occurred from 1587 to 1589, and it was especially severe in the vicinity of Roanoke Island.

The **Jamestown settlement** established in 1607 experienced great hardships and a high mortality rate in the first years. Tree ring analysis by Stahle et al. (1998) has indicated that Jamestown was started in the driest 7-year stretch (1606–1612) of the past *770* years.

II.A. TREE STUMPS AND LOGS (field trip)
MATERIALS

- field trip site that has several tree stumps or recently cut logs
- slices cut from tree trunks (if lab is held indoors)
- pieces of wood showing rings clearly, such as oak (piece of knotty lumber, wooden table, toy, carving, etc.)
- increment borer, if available (Ward's #10 W 1222; Forestry Suppliers, Inc. has a selection; or borrow from a nearby U.S. forest service, state forestry office, or cooperative extension office)

INVESTIGATIONS

☐ 1. Work in small groups of two or three to examine the cross sections of trees.

☐ 2. If the rings on the cut surface of a log or stump are difficult to read, use an increment borer to extract a core of wood.

Q5.　How old was your group's tree when it was cut? _____ (Estimate the age by counting its growth rings.)

Q6.　Determine which years had poor growing conditions and which had excellent growing conditions—express this in terms of the number of years before being cut if you do not know the cut date from which to count backwards. Look at the growth rings on the stump from the outside inward, recording which years had especially narrow rings (poor growth) or especially wide rings (excellent growth).
Poor year(s): _____
Excellent year(s): _____

9

□3. If you are walking through a cutover area, looking at stumps or at logs known to have come from this area, investigate the following questions.

Q7a. In which calendar year were the trees in this area cut? _____

Q7b. Compare your notes on especially narrow rings to the findings of students looking at other stumps in the area. Is there a consensus as to which years resulted in especially narrow growth rings?

Q7c. Do your years of poor growth correlate with droughts? Research the local weather records to see if there is a correlation.

□4. If you are examining disks or logs from an unknown source, investigate the following.

Q8a. Did any of the disks or logs have a whole series of consistently lopsided rings, all of which are wider on the same side of the tree? _____
This is called **reaction wood**; a leaning conifer tree or branch produces more wood on the underside, while a leaning hardwood tree or branch produces more wood on the top side.

Q8b. What would the ring pattern look like if a tree was vertical for the first 50 years but then, due to a landslide, it leaned for the next 50 years?

□5. Examine the cross-section portion of a piece of lumber, wooden tabletop, toy, or other wooden object and answer these questions.

Q9a. Count the rings on the cut end; how many years of growing effort went into making the wood in this piece?

Q9b. From which portion of the tree trunk was this piece cut? Draw it into the stem cross-section diagram of Figure 9.6 in Part II.B..

□6. Contemplate the following questions and answer them.

Q10. Suppose a tree grows for 50 years while quite heavily shaded by surrounding trees; then the surrounding older trees are cut, and this tree then grows another 20 years before being cut. Describe the tree ring pattern that you would expect to see on the cut stem, and tell why.

Q11. If all tree species in one corner of your state showed a very narrow growth ring for 1980, which environmental factor(s) would you suspect and which environmental factor(s) would you discount as being a possible cause for poor growth?

Q12. If only one of the trees in a grove had a narrow growth ring in 1980, what would you suspect to be the cause, and why?

Q13. Tropical trees in some areas grow at a constant rate throughout the year. What appearance would their wood have in cross section?

II.B. HOW DOES A TREE FUNCTION?
BACKGROUND

All that is needed for a tree to live is for the roots and the photosynthetic leaves to be connected by a continuous strand of three things (see Figures 9.4 and 9.6):

1. Xylem—to conduct water and minerals to the leaves
2. Phloem—to conduct sugars from the leaves to the roots to provide them with energy and keep them alive
3. Vascular cambium between the xylem and phloem—to make more functional xylem and phloem to replace worn-out tissue

Xylem and phloem are the vascular tissues; any plant that has *vascular* tissues is called a "vascular plant."

Additional structural features of a tree (see Figures 9.4 and 9.6):

- **Old xylem** in the center of the tree no longer serves in conduction, but serves only as structural support and as a waste dump.
- **Heartwood** is the name for the old central wood when it is dark-stained in certain species of trees.
- **Sapwood** is the light-colored wood surrounding the heartwood; the outermost several growth rings are active in conduction of water and minerals upwards.
- **Vascular rays** are the thin "spokes" on a tree cross section radiating from the center of the tree; they serve to store food and to transport it laterally (inwardly and outwardly) to living cells of the young xylem and phloem.
- Beneath the bark surface are thin **periderm** layers (see Figure 9.4); each periderm layer includes the **cork cambium** and a layer of waterproofing **cork**, which it formed.

Hollow Trees

☐ 1. Find and examine a living but hollow rotting-out tree trunk or branch. Contemplate and answer the following questions.

Q14. How might the decay organisms have gained entry to the initially solid tree?

Q15. What part of this tree is missing and what is still there? How can this hollow tree still live?

Q16. What good is a *hollow living* tree to the ecosystem?

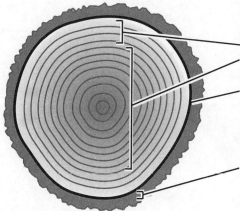

Xylem (wood)
Outer xylem is functional: conducts water and minerals upward.
Inner xylem is nonfunctional in conduction.

Vascular Cambium
Thin layer of growth (cell division)
- Makes new xylem to the inside
- Makes new phloem to the outside

Phloem (inner bark) transports sugars up and down.
Outer bark contains layers of cork cambium.
Cork cambium forms layers of waterproofing cork beneath the surface.

Figure 9.6 The regions of a stem and their functions

Q17. Some people believe that a properly cared for forest should have the *dead* trees removed. Why should dead trees be left in a natural forest ecosystem? List several points below.

How Does a Tree Heal Itself?

Study Figures 9.7 and 9.8 and notice how a tree gradually grows over a wound—a fire scar with dead vascular cambium or a pruned side branch. The vascular cambium at the edge of a wound forms new xylem to the inside as well as somewhat sideways over the wound area. This process continues year after year, slowly covering the wound. Also check out the striking front cover on the November 5, 1993 issue of the journal *Science* for a photograph of a cross section of an old giant sequoia stem that has experienced repeated fires followed by healing.

Knots (see Figure 9.9) are cross sections of side branches around which the main stem has grown. Knotholes in boards are **loose knots** that have fallen out; when a stem grows around a *dead* branch, the stem's new growth rings abut the dead bark on the dead branch. In such a board, there is no continuous growth of wood connecting the branch knot to the stem's wood; thus, the knot can fall out. The knots of knotty pine furniture do not fall out (**tight knots**) because those were the bases of *living* branches around which the stem grew; a growth ring of the stem was continuous with a growth ring of the living branch.

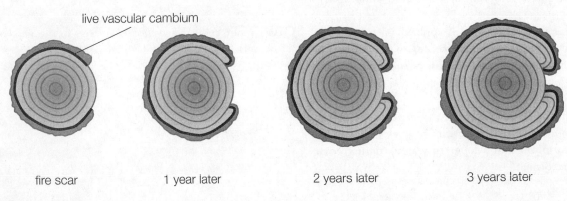

live vascular cambium

fire scar   1 year later   2 years later   3 years later

Figure 9.7 The process of healing over a scar

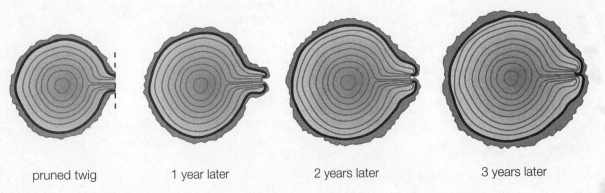

pruned twig   1 year later   2 years later   3 years later

Figure 9.8 The process of healing over a cut branch

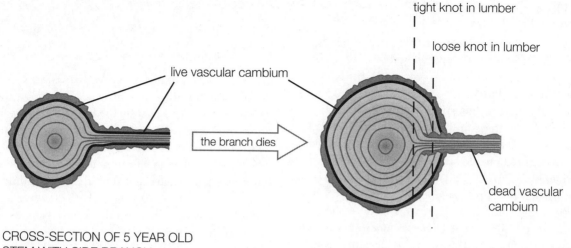

tight knot in lumber

loose knot in lumber

live vascular cambium

the branch dies

dead vascular cambium

CROSS-SECTION OF 5 YEAR OLD
STEM WITH SIDE BRANCH

3 YEARS AFTER THE BRANCH DIED

Figure 9.9 The process of engulfing a dead branch

☐2. Examine the stumps, logs, or stem slices of Part II.A for evidence of healed wounds.

Q18. Did you find places where the tree healed over a fire or mechanical scar, or engulfed a cut branch, nail, or barbed wire?

Skip ☐3. Evaluate the knots on any boards that you examined in Part II.A.

Q19. What kind of knots are they?

☐4. Look for trees on campus that have had side branches sawed off and that are in the process of healing. If you check them each year, you may notice the progress.

How Does a Tree Keep from Drying Out When Its Bark Keeps Ripping Open as It Grows Wider?

In the outer bark, thin layers of **cork cambium** keep forming, with the newest layers innermost. Each cork cambium layer produces a layer of waterproof **cork** to the outside. The cork cambium and cork layers are part of each of the **periderm** layers seen in the outer bark of Figure 9.4.

When the bark rips through the outmost corky periderm layer, there are already newer corky layers in place interior to that.

☐5. Look at the tree cross sections; can you see the thin corky periderm lines in the outer bark? Note that these lines are more easily seen in some species than in others.

☐6. Do all trees have rips in their bark, or does the bark growth of some tree species keep pace with the widening circumference of a growing tree? Keep an eye on the different barks when you walk past trees outside.

☐7. Do you find any tree with initials carved in the bark, or any saplings with wounds at their base from string trimmers?

Q20. How deep are these wounds as compared to the "growth rips" in the bark, and how might such wounds affect the health of a tree?

9

97

EXTENDED ASSIGNMENT

There is a great website on tree ring research—the Ultimate Tree-Ring web pages by Henri D. Grissino-Mayer of the University of Tennessee, Knoxville—at http://web.utk.edu/~grissino/. You can search the dendrochronology literature from that site and come up with such interesting things as the works of A. E. Douglass in the early 1900s on using dendrochronology to date old pueblo structures of the Southwest, or the paper by Burckle and Grissino-Mayer (2003) suggesting that the superior tone of the Stradivari violins is due to the wood having grown slowly and uniformly in the uniquely cool climatic conditions between 1645–1715. Go ahead! Search the literature from this site and find a research article that particularly interests you; read the article and summarize its contents in writing or orally, as instructed.

NOTE: Website addresses may change; if this website address has changed, do an Internet search for the organization, the person, or the name of the site. If this fails, search the bibliographic database BIOSIS using one or more keywords such as tree ring, climate, ecology, pollution, archaeology, dendrochronology, dendroclimatology, dendroarchaeology, and dendroecology.

PART III:
FORENSICS: WHAT CAN A SMALL STEM FRAGMENT REVEAL?

You are working in a forensics lab. There has been a break-in at an office downtown, and among the boot tracks left on the carpet by the burglar there was dried mud and some plant fragments that had dropped out of the boot treads. The fragments look like bits of some plant stem. You receive a prepared microscope slide of the specimen and are asked to identify the unknown.

Different groups of plants have different distinctive features in their internal stem structure that allow preliminary classification of an unknown. This can be accomplished through careful observation even *without* knowing stem anatomy in great detail. Further identification down to species by forensic botanists and wood specialists does require detailed knowledge of various cell types and their arrangement in stems.

The purpose of this investigation is for you to develop your observational skills and to realize that different types of plants have distinctly different internal stem structures.

OBJECTIVES

1. Practice critical observation, comparing and contrasting the stem structure of different species.
2. Be able to locate vascular bundles of a non-woody stem, and the xylem, phloem, and intervening vascular cambium of a woody stem.
3. Be able to determine if an unknown stem specimen matches one of a panel of known specimens solely on the basis of critical observation of the pattern of its internal microscopic structure.

MATERIALS

- prepared microscope slides of the following:
  - *Zea* (corn) stem, cross section (Carolina HT-30-3296; Ward's 91 W 7448)
  - *Medicago* (alfalfa) stem, cross section (Carolina HT-30-2780; Ward's 91 W 8012)
  - *Pinus* (pine), mature stem (showing two or more growth rings), cross section (Carolina HT-30-1298 or HT-30-5246; Ward's 91W 6312)
  - *Quercus* (oak), older stem, cross section (Carolina HT-30-5504)

Draw and record distinctive features of stem sections, observed microscopically.

corn	alfalfa
pine	oak
basswood	*Lycopodium*
bracken fern or maidenhair fern	The "unknown" is identified as _____ because of these internal structures:

9

- *Tilia* (basswood), 3-year-old stem, cross section (Carolina HT-30-2822 or HT-30-5528; Ward's 91W 8312)
- *Lycopodium* stem, cross section (Ward's 91 W 4624)
- *Pteridium* (bracken fern) rhizome, cross section (Carolina HT-29-9434; Ward's 91 W 5041) or *Adiantum* (maidenhair fern) rhizome, cross section (Ward's 91 W 4890)
- "unknown slide" provided by the instructor, label covered or removed
- compound microscopes, either one per student or set up with the prepared slides in position
- lens paper

INVESTIGATIONS

☐ 1. Examine microscopically each of the stem specimens provided; for instructions on using a compound microscope, see Lab 2 (Microscopy and Plant Cells). Determine what features are distinctive about each species as compared to all the others. Record your observations on the worksheet that precedes this page; you may write a precise description and/or make a clear drawing.

NOTE:

- Microscope slides of plant specimens are often stained with safranin O to stain thick, woody cell walls red. Therefore, you can often locate *xylem* cells by looking for large-diameter, thick-walled, red-stained cells. The thick woody walls of *fiber* cells will also stain red, but they can be distinguished from the conducting xylem cells because the cell walls of fibers are so extremely thick that there is almost no space left inside a fiber cell.
- It will help if you peek into your textbook to look at the figures in the section covering stems.

☐ 2. Acquire an "unknown" slide from the instructor and examine it carefully under the microscope.

☐ 3. Is the unknown slide clearly different from all the other specimens in internal structure, or is it like one of the identified specimens? Record your observations and conclusion on the worksheet.

Q21. If two plant species have very different arrangements of their internal stem structures, would you expect the two species to be more closely or more distantly related evolutionarily?

LITERATURE CITED

Botanical Society of America (BSA). 2001. "McIntosh Apple Development," http://mcintosh.botany.org/bsa/misc/mcintosh/mcintosh.html (accessed February 19, 2006). St. Louis, MO: Education Committee, Botanical Society of America.

Burckle, L., and H. D. Grissino-Mayer. 2003. Stradivari, Violins, Tree Rings, and the Maunder Minimum: A Hypothesis. *Dendrochronologia* 21(1):41–45.

Grissino-Mayer, H. D. 2006. "Ultimate Tree-Ring Web Pages," http://web.utk.edu/~grissino/ (accessed February 19, 2006). Knoxville, TN: Department of Geography, University of Tennessee.

Stahle, D. W., M. K. Cleaveland, D. B. Blanton, M. D. Therrell, and D. A. Gay. 1998. The Lost Colony and Jamestown Droughts. *Science* 280:564–567.

10

LAB

ROOTS

NOTES:

Today's lab is an example of **observational science** rather than experimental science. You will learn about roots through careful observation of specimens rather than through experimental testing of hypotheses. Much of our knowledge of plant structure and some plant processes comes from careful observations of many specimens over the years.

The objectives are given at the beginning of each part of this lab.

BACKGROUND

In general, if roots die, the plant dies. Roots are generally underground and out of sight, but they are critically important. Roots grow downward and typically anchor the plant in soil; from the soil, the roots absorb water and minerals that are then transported up the plant. The leaves use this water and minerals for various cellular functions including **photosynthesis**—the process that uses water, carbon dioxide, and solar energy to make energy-rich sugars and release oxygen. The sugars made by photosynthesis are then transported to other parts of the plant, such as down to the roots, to serve as an energy source there. In the roots, the sugars are broken down by **cellular respiration** in order to release their energy for use in various cellular processes—such as to drive the selective uptake of minerals into living root cells, completing the cycle.

The breakdown of sugars in the roots by cellular respiration requires oxygen and releases carbon dioxide as a waste product. The required oxygen is acquired by the roots from the air spaces between the soil particles. Long-term flooding can kill plants by depriving their roots of oxygen. Aquatic plants are adapted to growing in oxygen-poor waterlogged soils because they have evolved **aerenchyma** (Greek: *aer* = air, *en* = in, *chein* = to pour)—tissues with large air spaces that provide an avenue for oxygen diffusion from the aerial portions down to the roots. In Lab 11, "Leaves," you examine these air spaces in the leaves and leaf stalks of some aquatic plants.

The pioneers killed a large tree by **girdling**—the removal of a ring of bark from around the trunk. The sugars transported down to the roots travel through the phloem tissue of the inner bark; on a girdled tree, this sugar transport is blocked at the girdle. Once the sugar supply stored in the roots has been used up by cellular respiration, the roots have no energy source to keep the cells alive and to perform selective uptake of minerals. The roots die; and in turn, the tree dies.

Note that the molecular details of all the plant processes mentioned above are still active areas of research.

In this lab you will investigate the diversity of roots, the structure and function of the root tip, and the microscopic diversity of root cross sections.

NOTE: Your instructor may have you focus on only certain parts of this lab.

PART I:
DIVERSITY OF ROOTS

Roots come in a diversity of forms that allow their plants to survive in various habitats. Some roots look like stems, while some stems are mistaken for roots.

I.A. ROOT TYPES (field trip)

OBJECTIVES

1. Understand why roots underground perform cellular respiration.
2. Know the materials needed for cellular respiration in a cell.
3. Be able to identify the various kinds of modified roots.
4. Know how each kind of modified root functions to provide a survival advantage to its plant in its environment.

BACKGROUND

People generally underestimate the extent of root systems. It is often said that at least as much of a plant is underground as is aboveground; prairie plants can be 90 percent underground. Mentally, picture your campus; now mentally turn all the trees upside down; that's how much is under your feet.

There are two basic kinds of root systems (see Figure 10.1), although intermediate forms also exist:

- **Fibrous root system**—many roots are present, none of them predominating; most fibrous root systems are shallow to absorb the rainwater, but some can be quite deep.
- **Taproot system**—one long major root from which minor side roots arise.

There are also *modified* roots that serve not just in support and absorption, but in some additional function that promotes survival of the plant in its environment (see Figure 10.1):

Storage
- Carrot plant is a typical **biennial** (Latin: *bi* = two, *annus* = year); it stores food in the taproot in the first year, and then flowers and dies in the second.

Aeration
- Black mangroves grow on beaches; they are surrounded by many "pencils" protruding from the wet sand into the air; these are **pneumatophores** (Greek:

pneumatos = air, *pherein* = to bear) — vertical root outgrowths with aerenchyma and **lenticels**, which are tiny, often lentil-shaped breaks in the surface that provide places for gas exchange.

Additional types of support
- **Adventitious roots** are roots that form where they normally do not occur; they grow out of a plant part that is not another root; poison ivy stems produce abundant adventitious roots.
- **Prop roots** are adventitious roots that grow from a stem to help prop up the plant, such as corn or red mangrove.
- **Buttress roots** look like curtain folds flaring out at the base of tall rain forest trees growing in thin soils; buttress roots give the tree a broader base of support.
- **Aerial roots** is a general term referring to any roots that form in the air, including prop roots and various adventitious roots and true roots. The aerial roots of the banyan tree look like additional stems.

Propagation
- Buds are a standard feature on aerial shoots; however, **root sprouts** or **suckers** form when adventitious buds occasionally form on the roots of aspen, staghorn sumac, or tree-of-heaven; suckers from a single tree can produce a whole grove of trees.

Supplemental nutrient procurement
- Plants of the legume family such as clover and soybean have **root nodules**—bumps on their roots in which **nitrogen-fixing bacteria** live. These bacteria are one of very few organisms that can take the very stable and unusable nitrogen gas (N_2), break the two nitrogen atoms apart and add hydrogen atoms to each; the result is ammonia (NH_3), which is a useable form of nitrogen. Excess ammonia made by the bacteria is used by the host plant and some is released, fertilizing the soil. (In contrast, *root knots* are bumps resulting from an invasion of harmful microscopic roundworms called nematodes; you may have seen seed packets for "nematode-resistant" tomato varieties.)
- **Mycorrhizae** (Greek: *mykes* = fungus, *rhiza* = root) are fungi engaged in a mutually beneficial relationship with plant roots. The fungal strands penetrate a much greater volume of soil than the plant roots could reach. The plant receives some of the water and minerals absorbed by the fungus, while the fungus receives some of the photosynthetic products from the plant. The type of mycorrhiza called ectomycorrhiza causes the plant's root system to have many short, thick, side roots encased in whitish fungal strands.

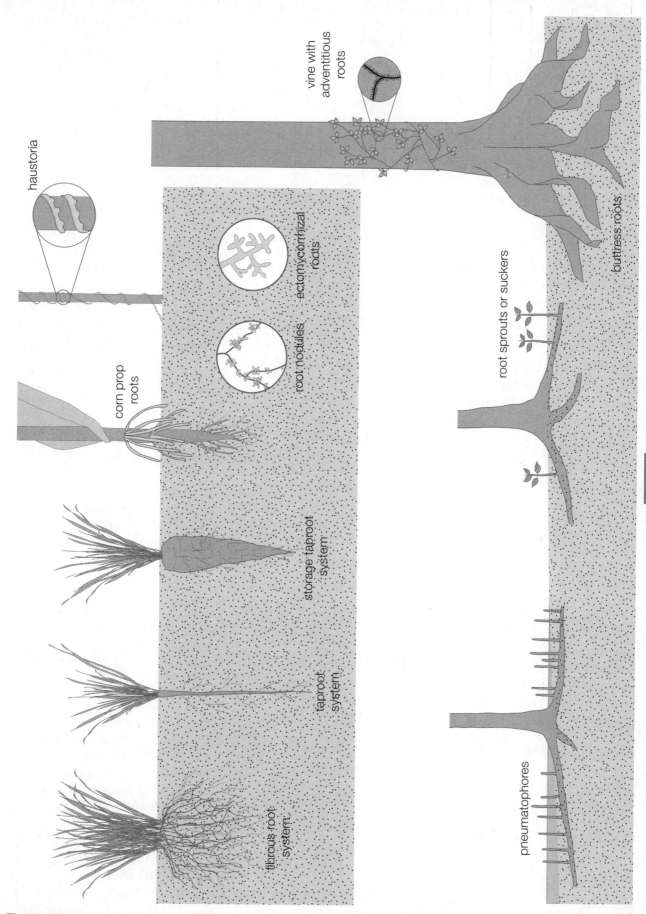

Figure 10.1 Root systems and root types

- Peg-like **haustoria** (Latin: *haurire, haustum* = to drink, draw) are produced by the twining stems of dodder, a parasitic non-photosynthetic flowering plant; the haustoria grow into the host plant to absorb nutrients.

MATERIALS

- trowels and spades
- buckets of water
- campus grounds and greenhouse

INVESTIGATIONS

☐1. If the soil outside is not frozen, your instructor will take you to a nearby location where you can dig up some "weeds." Dig deeply and try to get the entire root system of each plant. Rinse the root system in water to remove the soil particles.

Q1. Do all "weeds" have the same kind and same depth of root system? Describe the diversity of root systems found by the class. How many had fibrous roots, a taproot, adventitious roots, root nodules?

Q2. Which of the plants that you dug up had root nodules? _____

Q3. In ecology, the phrase "resource partitioning" refers to the situation in which various species can coexist in an area by using the resources differently and thus reducing competition among the species. Describe resource partitioning by root systems in a plant community.

☐2. Tour the campus and greenhouse, examining and identifying various types of roots. For those root types not seen on the tour, look at images in your textbook; or do an image search using Google®, for example.

☐3. Discuss in your group the adaptive advantages of each of the types of roots seen. Then do the "match it" exercise.

Q4. A neighbor purchased a decorative sumac with deep-toothed leaves and planted it in her large flowerbed. Each year thereafter, she has found new saplings of the same variety nearby; she yanks them out, but they keep on reappearing everywhere. The plant has never flowered and set seed. What is going on here?

MATCH IT!	
Adaptive advantages to the plants	**The structures**
____ 1. Provides abundant energy for rapid, vigorous growth immediately in the next growing season ____ 2. Retrieves minerals from deep in the ground where fewer plants have their roots ____ 3. Excellent at erosion control; has an abundance of root tips for absorption ____ 4. Form a broad base that helps to keep a tall tree from tipping over ____ 5. Enable a weak-stemmed green plant to be vertically oriented ____ 6. Enable a strong-stemmed plant having an inadequate true root system to be stably supported ____ 7. Enable one tree of a pioneer species to efficiently cover an open area with a whole grove of trees without first producing seeds ____ 8. Allow plants to grow in poor soils lacking various nutrients ____ 9. Allow plants to grow in nitrogen-poor soils ____ 10. Allow plants to survive in spite of lacking photosynthetic machinery ____ 11. Allow waterlogged roots to receive oxygen for cellular respiration	A. adventitious roots B. buttress roots C. fibrous root system D. haustoria E. mycorrhizae F. pneumatophores G. prop roots H. root nodules I. root sprouts, suckers J. storage root K. taproot

I.B. ROOT OR STEM?

BACKGROUND

Unlike stems, a root does not have leaves and lateral buds in the leaf axils (angle formed by the leaf stalk and twig). Although roots themselves do not produce leaves, the roots of certain species can occasionally produce an adventitious bud that grows into a root sprout, or sucker, that grows out of the ground and bears leaves (see Part I.A).

Unlike roots, a stem has nodes; a **node** is the place along a *stem* where a leaf or its evolutionary remnant is or was attached. On a leafless stem, a node can be located by looking for a leaf scar or leaf remnant that may look like a thin line on the stem or a line encircling the stem. The lateral bud at a node may also be large enough to be noticed. Stems have nodes, while roots do not.

You may see catalogs selling a variety of flower "bulbs"—variously shaped fleshy plant structures that have stored energy to fuel the rapid growth of aerial shoots and flowers in the spring. Some people mistakenly think of all these "bulbs" as roots, because they are buried in the ground. They are not roots, and only certain ones are actually bulbs. The botanical term **bulb** applies only to structures such as onions, consisting of a very short stem bearing a cluster of numerous thick, fleshy, non-photosynthetic leaves (see Figure 10.2). Thus, a bulb is just a large bud on a short stem. In contrast, a **corm** is a compact, short storage stem with some thin, papery leaves surrounding the outside; if you cut a corm in half, it is solid tissue rather than separate, thick, fleshy layers as in a bulb. A **rhizome** is a thick, horizontal, underground stem; being a stem, the rhizome has markings on it indicating the nodes. A **tuber** is a massively swollen region of underground stem; a tuber has numerous areas of lateral buds, the nodes or **eyes**, from which the sprouts grow; each eye consists of a leaf remnant (the crescent-shaped line), above which are the buds. Bulbs and corms have a basal plate at the bottom, but rhizomes and tubers do not.

MATERIALS

Labeled plant specimens are at stations around the room.

- white potato, sweet potato, ginger root, kohlrabi, radish, turnip, and onion
- flower "bulbs"—crocus or gladiolus; tulip or daffodil; iris
- paring knives

INVESTIGATION

☐ 1. Work in pairs. After studying the background information, examine and evaluate the vegetables provided. Determine for each vegetable whether it is a storage root or one of the kinds of modified stems. Record the diagnostic features observed, and identify the type of structure in Table 10.1.

☐ 2. Evaluate the specimens that are often called "bulbs." Look for key features and decide what kind of structure it is—bulb, corm, rhizome, or tuber. Record your observations and verdict in Table 10.1.

bulb

rhizome

corm

tuber

Figure 10.2 Bulbs and other fleshy underground structures inaccurately called "bulbs" in some catalogs

☐3. After the lab, the so-called ginger root can be planted in soil; it will grow into a fine flowering plant, and if you dig it up after a year, you will have even more ginger than you started with.

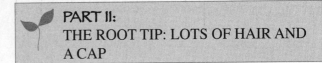

PART II:
THE ROOT TIP: LOTS OF HAIR AND A CAP

OBJECTIVES

1. Understand the function and importance of root hairs.
2. Observe the location, density, and extent of root hairs at root tips of various seedlings.
3. Understand the function and importance of the root cap.

BACKGROUND

For the typical plant to stay alive, its root system must absorb water and minerals. Absorption takes place mainly at the root hairs, which occur in a zone near each of the many root tips in the root system (see Figure 10.3). A **root hair** consists of a long finger-like outgrowth of an epidermal cell on the root. The many root hairs at a root tip dramatically increase the root's surface area that functions in absorption.

A root's **apical meristem** (Greek: *meristos* = divided, divisible) is the *region of cell division* at the root tip (see Figure 10.3). Behind this region of cell division is a *region of cell elongation*. The combination of cell division and subsequent elongation pushes the root tip through the soil. The epidermal cells that are in the *region of maturation* grow their root hair extensions into the soil.

Specimen	Diagnostic features observed	Type of root or stem
white potato		
sweet potato		
ginger root		
kohlrabi		
radish		
turnip		
onion		
crocus or gladiolus		
tulip or daffodil		
iris		

Table 10.1 Type of root or stem

The **root cap** protects the root tip as it pushes through the soil. The root cap is constantly being worn away by the rough soil and constantly regenerated by the apical meristem. Another critical role of the root cap is to detect gravity and cause the root to grow downward; the exact mechanism for this gravitropism is still being researched.

Books always seem to print photos of *radish* seedlings to show root hairs. Do all root tips look like those of radish? In this lab you will look for root hairs on radish as well as several other kinds of seedlings. You will observe the location, density, and length of root hairs at root tips and determine how much these features differ in the different species.

MATERIALS

- radish seeds (Carolina HT-15-9000, Ward's 86 W 8280)
- corn seeds (Carolina HT-15-9283, Ward's 86 W 8081)
- various other viable seeds (year-round in grocery stores: birdseed and bags of dried beans, peas, or lentils)
- paper towels or filter paper
- petri dishes
- foil
- dissecting microscopes
- dissecting needles

INVESTIGATIONS

☐1. Fold a paper towel into a square, and cut off the corners so that it fits into a petri dish. Add water to produce a wet, low platform that sits in a pool of water in the dish and wicks up water to keep the seeds on top uniformly moist throughout the experiment.

☐2. There should be no water standing on the top surface. (Note that when sopping wet, the root hairs

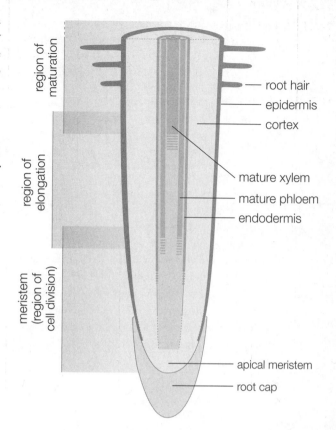

Figure 10.3 The root tip

will cling together just like human hairs, and thus they would not be very apparent.)

☐3. Evenly space a half dozen seeds on the wet towel in the petri dish. Cover with the lid, and wrap with foil.

☐4. Prepare one such dish each for radish, corn, and two other kinds of seeds.

☐5. Take these dishes home; check them each day. Seeds may take a few days to two weeks to germinate, grow, and fully develop their root hairs. In Table 10.2, record your observations on each species when their root hairs are well developed.

Species	How far from the end of the root does the root-hair zone start?	What length of the root is covered with root hairs?	How well developed are the root hairs (length, abundance)?

Table 10.2 Root hairs

Q5. Do all the seedlings have root hairs? _____ Can you see a reason for *radish* seedlings always being used to show root hairs in books? What is it?

Q6. Examine the root tips under a dissecting microscope; do you notice a gap between the tip of the root and the beginning of the root-hair zone (also see Figure 10.3)? New epidermal root cells complete elongation before they start to produce their root hairs. Why is it advantageous to a plant to grow root hairs into the soil *after* rather than before the time of elongation? (Consider what would happen to a root hair that grew out into the soil before its cell finished elongation.)

Q7. Explain why seedlings that are transplanted will generally be wilted for a few days and then look normal. (Consider what happens when you dig up the roots of a seedling for transplantation; consider that wilting results whenever the leaves lose more water than is being delivered to them.)

Q8. What other examples of efficient absorptive structures can you think of that have a large surface-to-volume ratio—structures that are themselves long and thin in shape (one is mentioned earlier in this lab) or that have finger-like projections? (Hint: Look up the human digestive tract in a biology book.)

PART III:
FORENSICS: WHAT CAN A ROOT FRAGMENT REVEAL?

OBJECTIVES

1. Practice critical observation, comparing and contrasting the root structure of different plants.
2. Be able to locate the cortex, endodermis, and large xylem cells in an herbaceous root.
3. Based on critical observation of the pattern of internal structures, be able to determine whether an unknown root specimen is similar in structure to any of an array of specimens that you have previously examined.

BACKGROUND

Roots of flowering plants do not all have the same internal structure. They *do* generally start out having an outer **cortex** region (storage tissue), then an **endodermis** layer (regulates the water and minerals that can pass into the interior), and then the **vascular tissue** (large, thick-walled **xylem** to conduct water and minerals upward, and thin-walled **phloem** to conduct sugars in either direction). The location, size, and extent of these structures, however, differ in the roots of different plant groups.

Figure 10.4a shows three common types of root cross sections: corn (a "typical" monocot), buttercup (traditionally called a "typical" herbaceous dicot), and basswood (a "typical" woody dicot); examine the pictures and note the unique distinguishing features in each. Figure 10.4b shows the stem sections of these plants; note that the stems have unique features that distinguish them from each other as well as from the roots. Forensic botanists well trained in plant anatomy can match or even identify small fragments of roots and stems.

In this lab, you will closely examine the root cross sections of several species to investigate the following questions: Are all monocot roots structured like a corn root? (Monocots include the grasses, greenbriers, lilies, orchids, and others; they tend to have flower parts in sets of three, and leaves with parallel veins, along with other features.) Are all dicot roots structured like the herbaceous buttercup or the woody basswood root? (Dicots include geraniums, sunflowers, alfalfa, willows, and basswoods; they tend to have flower parts in sets of four or five, and leaves with netted veins, along with other features.)

(a)

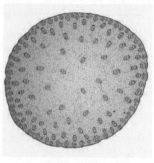

cortex · endodermis

xylem

monocot root (corn)

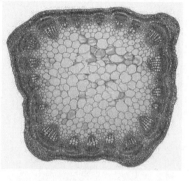

cortex

endodermis

xylem

herbaceous dicot root (buttercup)

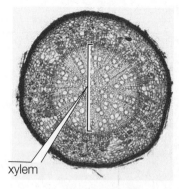

xylem

woody dicot root (basswood)

(b)

monocot stem (corn)

herbaceous dicot stem (alfalfa)

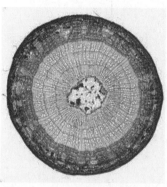

woody dicot stem (basswood)

10

Figure 10.4 Cross sections of (a) roots and (b) stems of monocot, herbaceous dicot, and woody dicot

MATERIALS

(Note that www.triarchmicroslides.com has a large selection of botanical slides)

- compound microscopes, 1 per student or set up with prepared slides in position
- lens paper
- *Zea mays* (corn) and *Ranunculus* (buttercup) root, c.s. (Carolina HT-30-1898, Ward's 91 W 9910)
- *Tilia* (basswood) root, cs (Carolina HT-30-2072, Ward's 91 W 8302
- *Smilax* (greenbrier) root, c.s. (Carolina HT-30-2480, Ward's 91 W 7361)
- *Helianthus* (sunflower) root, c.s. (Carolina HT-30-2150, Ward's 91 W 7912)
- *Medicago* (alfalfa) root, c.s. (Triarch 13-8)
- Orchid, aerial root, cs (Ward's 91 V 7300)
- 2 compound microscopes set up with 2 "unknown" roots slides, their labels covered with numbers

INVESTIGATIONS

☐1. In Figure 10.4, draw an arrow to the unique features in each cross section that helps you to distin-

guish it from the others.

☐2. For instructions on using a compound microscope, refer to Lab 2 (Microscopy and Plant Cells). Examine the cross sections of the roots of corn, buttercup, and basswood under the microscope; these sections should look essentially identical to the pictures in Figure 10.4.

☐3. Now examine the greenbrier, sunflower, alfalfa, and orchid root cross sections, comparing each to the roots in Figure 10.4. Draw each cross section on Worksheet 1; if a root's structure is distinguishable from the others, describe and draw arrows to the features that are distinctive; if the root's structure is identical to another species observed, indicate that conclusion.

☐4. Examine the two "unknown" roots on numbered slides; an unknown may be one of the species that you have already examined or may be something entirely different. On Worksheet 1, identify the unknown, indicating the structures that allowed you to identify it, or draw the unknown if it is something entirely different.

greenbrier root	sunflower root
alfalfa root	orchid root
unknown root 1 = _____	unknown root 2 = _____

PART IV:
AERENCHYMA IN AQUATIC PLANTS

Emergent plants have their lower portions submerged in water. The plant's roots, as well as a portion of the stem or the whole rhizome (horizontal underground stem), grow in the oxygen-poor mud. These plant parts do require oxygen for cellular respiration, but they generally cannot acquire adequate amounts from the surrounding mud. These plants contain aerenchyma tissue that provides continuous passageways for oxygen movement from the leaves down to the roots.

Many flooding-resistant plants respond to submergence by rapidly producing an abundance of adventitious roots with well-developed aerenchyma (see Blom et al. 1994 and Seago et al. 2005, as well as other papers in the September 2005 special issue of *Annals of Botany*, which is devoted to the topic of flooding stress).

In this part of the lab you will discover an emergent plant's internal structure, which provides internal aeration for its submerged plant parts.

OBJECTIVES

1. Understand why roots need oxygen.
2. Understand the structure of the aerenchyma of an emergent aquatic plant that provides an avenue for oxygen to reach the roots.

MATERIALS

- compound microscope per student
- lens paper, microscope slides and cover slips, and dropper bottles of water
- single-edge razor blades
- living cattail plant (*Typha*) or another emergent aquatic plant in your area (sedges, rushes, sweet flag, etc.)

INVESTIGATION

☐1. Carefully make paper-thin cross sections at intervals down the leaf, stem, and roots of the emergent aquatic plant. Place each section in a drop of water on a microscope slide; cover with a cover slip; examine under the compound microscope. For instructions on using a compound microscope, refer to Lab 2 (Microscopy and Plant Cells).

☐2. Draw the cross sections of the various parts of your plant, labeling the air spaces (Worksheet 2).

Q9. What is the source of the oxygen that then travels from the leaves down the air spaces in the aerenchyma to the roots?

EXTENDED ASSIGNMENT

At gardening centers and in seed catalogs, you sometimes see preparations of *Rhizobium* (nitrogen-fixing bacteria) that are to be sprinkled in the soil along with your pea and bean seeds. Burpee's catalog, for example, lists "Burpee Booster for Beans and Peas." How much of an effect will this preparation have on the growth of the pea and bean plants? The class can design an experiment to test this. You could use sterile sand as the growing medium. Remember that the characteristics of a properly designed scientific experiment are replication, a control, and a single variable. Don't forget to make arrangements for the regular watering of the pots.

LITERATURE CITED

Blom, C. W. P. M., A. C. J. Voesenek, M. Banga, W. M. H. G. Engelaar, J. H. G. M. Rijnders, H. M. Van de Steeg, and E. J. W. Visser. 1994. Physiological Ecology of Riverside Species: Adaptive Responses of Plants to Submergence. *Annals of Botany* 74:253–263.

Seago, J. L. Jr., L. C. Marsh, K. J. Stevens, A. Soukup, O. Votrubová, and D. E. Enstone. 2005. A Reexamination of the Root Cortex in Wetland Flowering Plants with Respect to Aerenchyma. *Annals of Botany* 96:565–579.

10

11

LEAVES

LAB

I. What's in a leaf, and why?
II. Surviving the extremes—structural adaptations
III. The leaf surface—telltale clues
IV. Weird leaves—modified for special functions

Plant leaves share basic design features, but they differ in detail from species to species. A plant's microscopic leaf structure is genetically determined, differing according to:

- the plant group (lilac and pine differ, as seen in Part I),
- the type of photosynthesis performed by the plant (grasses with C3 and C4 photosynthesis differ),
- the type of habitat in which the plant occurs (desert and aquatic plants differ, as seen in Part II). The hairs, glands, and cell shape on the leaf surface (Part III.A) can be distinctive enough for forensic botanists to match or even identify small leaf fragments.

The structure of leaves on one plant, although genetically determined, can vary somewhat due to environmental conditions—whether the leaf is in the sun or in the shade, for example (Part II). In addition, there is some plasticity (variation) in the arrangement of the leaf pores, for example (Part III.B).

Some leaves are so highly modified for special functions that they do not even look like leaves, while some modified stems look like leaves (Part IV).

This lab will develop your observational skills, by comparing and contrasting leaf structures. It will also develop your ability to prepare finely sliced specimens and epidermal peels.

 PART I:
WHAT'S IN A LEAF, AND WHY?

OBJECTIVES

1. Know the basics of photosynthesis—the chief function of leaves.
2. Know the various structures in a leaf of a broad-leaved mesophytic plant.
3. Know how each leaf structure promotes photosynthesis.
4. Recognize the structural differences in the leaves of a eudicot (lilac), of a monocot (corn), and of a conifer (pine).

BACKGROUND

Green leaves come in a great diversity of shapes and sizes; these diverse leaves are, however, very similar in nearly always being thin and in sharing common features of their microscopic cellular structure. Why this similarity in the structure of leaves? It all comes down to their main function, photosynthesis.

Photosynthesis (Greek: *photo* = light) is the process by which green plants get their energy from light. Leaf cells that are photosynthetic contain numerous small green organelles called **chloroplasts** (Greek: *chloros* = green, greenish-yellow, *plastos* = formed) in which the photosynthetic process occurs. In the presence of the green **chlorophyll** (Greek: *phyllon* = leaf) pigment, photosynthesis uses carbon dioxide and water to convert solar energy into the chemical bond energy of glucose sugars; oxygen and water are released as by-products.

Leaf structure has evolved to optimize photosynthesis. Leaves usually are *thin* to permit light penetration to all of the internal photosynthetic cells; leaf *pores* can

NOTE: Your instructor may have you focus only on certain parts of this lab.

be opened to admit carbon dioxide and release oxygen; leaves have *conducting tissue* in veins to transport the water to the cells and sugars to other parts of the plant as needed. Figure 11.1 shows the structures in a typical leaf from a broad-leaved plant (lilac) that is a **mesophyte** (Greek: *mesos* = middle, *phyton* = plant)—a plant from areas of moderate moisture.

In this part of the lab, you will examine various prepared and fresh slices and peels of leaves; you will then use your observations to deduce the functions of the various leaf structures. You will also see the diversity in the basic leaf structure by examining leaves of different plant groups: lilac, corn, and conifer.

MATERIALS

* microscope materials: compound microscope per student, lens paper, microscope slides and cover slips, and dropper bottles of water
* new single-edge razor blades
* permanent slides of leaf cross sections (Note that www.triarchmicroslides.com has a large selection of botanical slides.)
  - *Syringa* (lilac) (Carolina HT-30-3790, Ward's 91 W 8272)
  - *Zea* (corn) (Carolina HT-30-4054, Ward's 91 W 7454)

- Pine, five-needle type, white pine (Ward's 91 W 6503, Triarch 10-6F)
* pine branch with bundles of needles
* living leaves of a mesophyte: privet, maple, sunflower, or another
* living shoots of Swedish ivy or another leaf that peels easily

INVESTIGATIONS

You may be instructed to work in groups.

☐1. Following the instructions for using a compound microscope (in Lab 2, Microscopy and Plant Cells), examine the permanent slide of a lilac leaf, and find the structures indicated in Figure 11.1. Note that the photosynthetic cells contain little green pill-shaped chloroplast organelles in which photosynthesis occurs. However, also note that the colors seen in permanent slides are not true, because such slides are artificially colored with various stains to increase contrast. Typically, regular cell walls and cytoplasm are stained green, nuclei and heavy cell walls are stained red, starch grains are stained purple, and the chloroplasts may look like numerous reddish spots within cells.

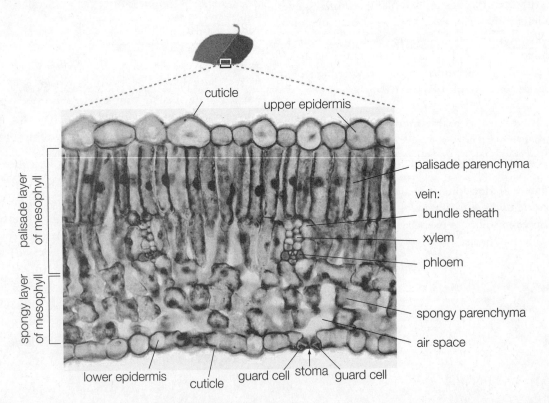

Figure 11.1 The parts of a typical leaf blade of a broad-leaved mesophyte (lilac). The staining in this preserved specimen obscures the fact that the epidermis is clear.

Q1. Are the stomata on the upper and/or lower epidermis of lilac leaves? _____Why is this the most advantageous position? (Consider moisture loss from a sunny versus a shady surface.)

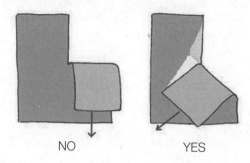

NO   YES

Figure 11.2 Tearing a leaf for an epidermal peel

☐2. To see leaf cells in natural color, make your own slices of the leaf blade of a mesophyte. Using a sharp razor blade, make several *paper-thin* slices and slide them, on their sides, from the razor blade into a drop of water on a microscope slide; cover with a cover slip.

☐3. Examine your fresh mount with the compound microscope; locate the green chloroplast-containing photosynthetic cells and the clear non-photosynthetic cells.

Q2. Which cells of a leaf are non-photosynthetic?

☐4. Prepare a peel of the *lower* epidermis of Swedish ivy by tearing *across*, as shown in Figure 11.2, rather than tearing straight down. (Remember that under high power, only a tiny area of a specimen is viewed; a peeled epidermal strip that is only 2 mm wide is large enough.) Lay the leaf edge with the attached peel into a drop of water on a microscope slide; use a razor blade to detach the peel from the leaf; cover the peel with a cover slip.

☐5. Carefully examine the epidermal peel and find a clear area that has no underlying mesophyll cells attached to the epidermis. Locate the **stomata** (Greek: *stoma* = mouth), each surrounded by a pair of guard cells. Note that in a wet mount of a peel, some stomata may trap air bubbles above or beneath them, producing an odd appearance.

11

MATCH IT!	
Leaf Structures	**Functions**
____ 1. cuticle	A. Pair of cells that use energy to change their shape, resulting in opening or closing the stoma as needed
____ 2. epidermis	B. Photosynthetic cells (true for four of the listed structures)
____ 3. palisade layer of mesophyll	C. Pores through which carbon dioxide enters the leaf and oxygen leaves the leaf (in other words, pores for gas exchange)
____ 4. spongy layer of mesophyll	D. For transporting water, minerals, and sugars; the plumbing of the cell
____ 5. bundle sheath	E. These provide avenues for gases to move freely throughout the interior of the leaf and reach every photosynthetic cell
____ 6. vein or vascular bundle	F. Tough, compact, translucent cells that protect the inner leaf tissues, provide support, and hold water
____ 7. air spaces	G. Waxy barrier that helps retard water loss through the leaf surface, but also prevents gas exchange
____ 8. stomata (singular: stoma)	
____ , ____ 9. guard cells	

Q3. When viewed under high power, do any of the cells in the lower epidermis of Swedish ivy contain green chloroplasts? _____
Which cells? _____
What is the function of these cells? Why would they in particular benefit from containing the energy-producing photosynthetic chloroplasts?

Q4. Considering the location of the structures shown in Figure 11.1, and using your own observations and your logic, do the "Match It" exercise.

☐6. The arrangement of cells in the interior of the leaf varies with plant groups. Consult Figure 11.3 as you examine the prepared leaf cross sections of two flowering plants (lilac and corn) and a conifer (5-needled pine). Record your observations in Worksheet 1.

☐7. Look at the pine branch and note how the needles are in bundles. Look at the needle cross section of a two-needled pine in Figure 11.3, and note that the cross section is shaped like half of a circle. Now note the shape of the needle cross section of the five-needled pine on your microscope slide—if you move all five cross sections together, you form a circle.

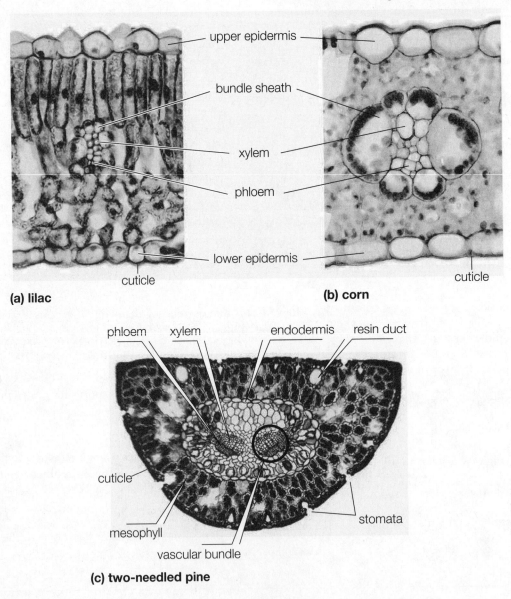

(a) lilac

(b) corn

(c) two-needled pine

Figure 11.3 Comparison of leaf of lilac, corn, and two-needled pine

116

	Lilac Leaf	Corn Leaf	Pine Leaf
Has noticeably thick cuticle compared to the other two species			
Has sunken stomata; guard cells are not even with the leaf surface			
Has the mesophyll divided into palisade and spongy layers			
Has a wreath of mesophyll cells tightly surrounding the bundle-sheath cells			
Has ducts lined with resin-secreting cells; the resin will seal any wounds			
Has an endodermis layer that regulates which materials reach the veins			

Q5. What shape would you expect the needle cross section to be for a pine with three-needled bundles?

Q6. Plants with the special C4 photosynthesis process have a tight wreath of mesophyll cells surrounding the bundle-sheath wreath around a leaf vein. Based on the leaf structure, which has C4 photosynthesis: lilac, corn, or pine? _____

PART II:
SURVIVING THE EXTREMES—STRUCTURAL ADAPTATIONS

OBJECTIVES

1. Be able to identify the adaptive structures on a xerophytic leaf and tell how each promotes survival.
2. Be able to identify adaptive structures on floating and on submerged leaves of hydrophytes, and tell how each promotes survival.
3. Be able to evaluate an unidentified leaf and deduce whether it is from a xerophyte, a mesophyte, or a hydrophyte.
4. Be aware of plasticity in plant structure and be able to distinguish a leaf growing in the sun and one in the shade on the same plant.
5. Know the structure and function of the abscission zone that is formed before a leaf drops.

11

BACKGROUND

A leaf's thinness is optimal for photosynthesis but results in a very large surface area through which moisture can be lost. The waxy **cuticle** (Latin: *cuticula* = little skin) blocks much of the moisture loss through this surface; however, the stomata, when open to allow gas exchange for photosynthesis, are major avenues for moisture loss from the interior of the leaf. More than 90 percent of the water lost from a plant is lost through stomata, in a process called **transpiration**. It is the water released by transpiration that condenses on the inner walls of a closed terrarium. Some of the moisture loss from the leaf's internal surfaces is beneficial in that it produces **evaporative cooling**. However, excessive moisture loss from leaves must be prevented; this is especially critical for **xerophytes**, plants of dry areas (Greek: *xeros* = dry).

Compared to mesophytes, xerophytes such as oleander and grasses (Figure 11.4) have various additional structural and physiological adaptations to retain leaf moisture. Key xerophytic strategies are:

1. Have more protection between the dry outside air and the photosynthetic cells (thicker cuticle; multiple epidermis).
2. Have cells dedicated to water storage (cells inside the leaf; multiple epidermis).
3. Reduce the transpiration rate by reducing the rate of air flow past the stomatal openings (hairy leaf surface; curled-under leaf edges; stomata that are not even with the leaf surface, but rather sunken in some manner; grass epidermis with clusters of large **bulliform cells** (Latin: *bulla* = bubble) also called **motor cells** that, upon losing water, cause the leaf to roll or fold up).
4. Have special C4 or CAM photosynthesis that is more efficient at binding carbon dioxide, thus reducing or eliminating the need for stomata to be open during the heat of the day. With fewer stomata open, there is less moisture loss.

Hydrophytes (Greek: *hydro* = water) are plants that grow in wet areas, either in mud or partially to wholly submerged in water. *Floating* and *emergent* leaves have blades and stalks with **aerenchyma**, tissue with large air spaces (Greek: *aer* = air; *en* = in; *chein* = to pour). The aerenchyma helps in floatation as well as provides air passageways to the submerged roots for gas exchange; the submerged roots get energy by breaking down sugars in **cellular respiration**, a process that requires oxygen and releases carbon dioxide. In totally *submerged* leaves, very thin ribbonlike or threadlike leaf blades are advantageous in that water and dissolved gases can readily reach all cells in such leaves. On the other hand, stomata, cuticle, and xylem tissue for conducting water up a plant would give no survival advantage to a submerged hydrophyte and thus are missing.

Any genetic change (mutation) that eliminates an unneeded feature or results in a beneficial new feature makes the plant better adapted to its environment; that plant is more likely to survive and pass that mutation on to the next generation.

Some plants experience seasonal extremes—desert dry seasons or cold, snowy winters (when precipitation is tied up as snowpack). Many such plants are **deciduous** (Latin: *decidere* = to fall off)—they drop all of their leaves before the extreme season and reduce their metabolism to a minimum. Other plants survive such extreme seasons by producing xerophytic leaves that are retained.

In this part of the lab, you will examine adaptive features on the leaves of xerophytes and hydrophytes, and then use your knowledge to identify the habitat of unknown specimens. You will also examine sun and shade leaves to investigate the environmentally induced variation on a plant. Finally, you will examine the **abscission zone** that forms across the base of the leaf stalk to facilitate leaf drop.

MATERIALS

NOTE: At www.triarchmicroslides.com you'll find a large selection of botanical slides; you can do a search there for "xerophytic leaf," for example, and get a listing of all such slides available.

- microscope materials: compound microscope per student, lens paper, microscope slides and cover slips, and dropper bottles of water
- *Nerium* (oleander) leaf, c.s. (Carolina HT-30-3826, Ward's 91 W 8052)
- *Elodea* leaf, c.s. (Carolina HT-30-4066, Ward's 91 W 7128)
- "unknowns" with labels covered, set up on compound microscopes:
  - *Ficus elastica* leaf, c.s. (Carolina HT-30-3862, Ward's 91 W 7902)
  - *Ammophila* (beachgrass) leaf, c.s. (Carolina HT-30-4042, Ward's 91 W 7071)
  - Yucca leaf, c.s. (Carolina HT-30-4114, Ward's 91 W 7421)
  - *Vaccinium macrocarpon* (cranberry) leaf, c.s. (Triarch 15-383)
  - *Ligustrum* (privet) leaf, c.s. (Carolina HT-30-3838, Ward's 91 W 7940)
- new, single-edge razor blades
- living plants with aerenchyma: water-lily leaf and petiole, water hyacinth, and cattail leaf
- *Nymphaea* (water-lily) leaf, c.s. (Carolina HT-30-3892, Ward's 91 W 8061)
- sun and shade leaves, c.s. (Carolina HT-30-3556, Ward's 91 W 9930)
- *Acer* (maple) leaf abscission, median l.s. (Triarch 15-1SP1*)
- Internet connection

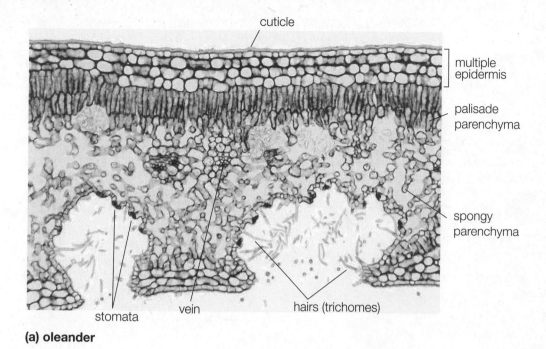

cuticle

multiple epidermis

palisade parenchyma

spongy parenchyma

stomata   vein   hairs (trichomes)

(a) oleander

turgid bulliform cells

flaccid bulliform cells

(b) grass

Figure 11.4 Oleander and grass leaves have structural features to reduce moisture loss.

INVESTIGATIONS

Work in groups.

☐1. Xerophyte. Examine the slide of the xerophytic oleander leaf. In Worksheet 2, note the form of each leaf structure, and describe how each form possessed by the oleander leaf promotes survival in dry areas.

Q7. Why do pine leaves (needles) have sunken stomata? Explain. (Consider the climatic conditions in which conifers usually exist.)

Q8. Barefooted children in summer learn that clover (C3 plant) is cooler to the feet than various grasses (C4 plants). Reread the background section and suggest an explanation for this.

☐2. Hydrophyte. Examine the prepared leaf cross section of *Elodea*, a submerged hydrophyte, under the microscope. On Worksheet 3, note the structures of *Elodea*, and describe how each structure makes the leaf well adapted to the environment.

WORKSHEET 2. Oleander

Leaf structure, and its possible forms	The form of this structure in oleander	Advantage of this form to a xerophyte
Cuticle: not apparent, thin as in lilac, or thick		
Epidermis: missing, one cell layer, or multiple cell layers		
Stomata: missing, even with epidermis, sunken, or in stomatal crypts (depressions)		
Area around stomata: hairless or hairy		
Water-storage cells: absent or present		
Leaf margins: flat or curled-under		
Bulliform cells: absent or present		

WORKSHEET 3. *Elodea* (submerged hydrophyte)

Structure, and its possible forms	The form of this structure in *Elodea*	Advantage of this form to a submerged hydrophyte
Cuticle: not apparent, thin as in lilac, or thick		
Epidermis: missing, one cell layer, or multiple cell layers		
Stomata: missing, even with epidermis, sunken, or in stomatal crypts (depressions)		
Veins: normal or poorly developed with little thick-walled xylem		
Thickness of leaf blade: 2, 4, or more cell layers		

☐3. Aerenchyma. Make paper-thin sections of the leaf blade and petiole (leaf stalk) of the plants provided: water-lily leaf and petiole, water hyacinth petiole, and cattail leaf. Mount the sections and examine under the compound microscope; draw the cross section in Worksheet 4, labeling the air spaces.

Q9. Water marigold has threadlike leaves on the lower portion of the stem and broad, flat leaves on the upper portion. Why might this be so? What does this tell you about its environment?

Q10. On which side of a floating leaf blade (water-lily) are the stomata located? Why are the stomata located there? (If you cannot see the stomata in your hand section, then closely examine the prepared slide of the water-lily leaf.)

☐4. Unknowns. Now it's your turn! You have seen examples of mesophytic, xerophytic, and hydrophytic leaves. Examine the leaf sections of several "unknown" plants set up on microscopes. For each leaf specimen, record on Worksheet 5 the structural features and deduce whether the leaf is a mesophyte, a xerophyte, or a hydrophyte.

☐5. Sun and shade. Examine the slide of the cross sections of sun and shade leaves. Leaves in the shade are often larger and thinner than leaves that are in the sun on the same plant.

Q11. What structure is thicker or has more layers in the sun leaf than in the shade leaf?

Q12. Why would it be an advantage for the sun leaf in bright light to have this structure thicker? Why would it be advantageous for the shade leaf in dim light to be thinner and broader?

☐6. Leaf drop. Look at the slide of the abscission zone—the zone of weakness across the base of the leaf stalk where the leaf will break off the plant. Look for two layers in the zone—the **separation layer** (the layer of weak cells where the break will occur) and the **protective layer** (waterproof layer of corky cells).

Q13. After leaf drop, would you expect the corky protective layer to be on the end of the leaf stalk or on the twig? _____

Q14. What is the benefit to the plant of having this protective layer? What would happen if there were none? (*Hint*: Think of a scab.)

Q15. How does a leaf scar on a twig (see Figure 9.1 in the "Tree Investigations" lab) relate to the abscission zone?

11

WORKSHEET 4. Aerenchyma

water-lily leaf	water-lily petiole	water hyacinth petiole	cattail leaf

Feature	Specimen 1	Specimen 2	Specimen 3	Specimen 4	Specimen 5
Cuticle: none, thin, thick					
Layers of epidermis: none, one, multiple					
Stomata position: even, sunken, crypt					
Hairiness of stomatal area					
Leaf edges: flat, curled under					
Water-storing cells: absent, abundant					
Bulliform cells: absent, present					
Thickness of leaf blade: 2, 4, or more cell layers					
Xylem in veins: poor or present					
Conclusion: meso-, xero- or hydrophyte					

Q16. When you prune a hedge in summer, the leaves on a cut-off twig will dry up but remain firmly attached; you have to use force to rip those leaves off, whereas a simple tap will make an autumn leaf fall off. Explain.

water pressure of bulliform cells that cause grass leaves to fold or curl up). At dawn, the leaves reorient to be horizontal. To view these "sleep movements," go to the Plants-In-Motion website by Roger Hangarter of Indiana University, at http://plantsinmotion.bio.indiana.edu/plantmotion/starthere.html. Click on *Circadian Responses* and view the time-lapse videos (Latin: *circa* = about, *dies* = day). While you are there, also click on *Nastic Movements* and view other videos of nondirectional movements in response to stimuli. (If the web address has changed, do a search for the producer or name of the website.)

Q17. What benefits might a plant derive from having vertically-oriented leaf blades at night?

☐7. Day/night. The change from light to dark in the 24-hour cycle triggers some plants to change the orientation of their leaves from horizontal to vertical. This movement is due to water moving into and out of certain cells, causing changes in internal water pressure (remember the changes in internal

PART III:
THE LEAF SURFACE— TELLTALE CLUES

1. Become familiar with the diversity in the shapes of hairs, glands, and epidermal cells, and in the stomatal arrangements.
2. Practice carefully noting distinctive structures on a specimen.
3. Apply your knowledge of previously observed specimens to identify "unknown" specimens.
4. Become aware that stomatal arrangement is genetically determined but also somewhat variable (Part III.B).
5. Practice designing and conducting your own investigation of a question (Part III. B).

III.A. HAIRS, GLANDS, AND CELL SHAPE

BACKGROUND

A variety of distinctive external features, along with the distinctive internal structures covered in Parts I and II, assist the forensic botanist in identifying leaf fragments on a suspect, or in stomach contents of a victim. (Read about some case histories in the 1990 *BioScience* article "Forensic Botany," by M. Lane et al.) Epidermal hairs may be uni- or multicellular, straight or wavy or branched, stiff or soft, long or short, and visible to the naked eye or only under a microscope. Glands may be at the epidermal surface (Swedish ivy) or at the end of a glandular hair (tobacco). Epidermal cells may be shaped like bricks, polygons, or highly convoluted jigsaw-puzzle pieces. The stomatal arrangement in the epidermis may be scattered, grouped, or in lines. The guard cells may be shaped like kidney

beans or dumbbells. Each pair of guard cells may be surrounded in a precise pattern by two or more distinctly different cells (**subsidiary cells**).

In this part of the lab, you will characterize the surface features of several leaves, and then identify an unknown specimen as being one of these species or something different.

MATERIALS

- microscope materials: compound microscope per student, lens paper, microscope slides and cover slips, and dropper bottles of water
- new single-edge razor blades
- bottles of clear fingernail polish
- leafy vegetables: lettuce, spinach, and onion
- *Begonia*, Swedish ivy, and *Zebrina*
- a variety of other locally available leaves from which students can choose, such as mullein, Russian olive, oak, sycamore, tomato, lavender, tobacco, petunia, grape, geranium, *Coleus*, *Sedum*, and *Dianthus*
- microscope with a mounted "unknown" epidermis (identical to one of the specimens provided or a totally new one)

INVESTIGATIONS

Work in groups.

☐1. Make lower epidermal peels of lettuce, spinach, and onion (the thick layers on an onion are fleshy leaves). If you cannot get a peel, then get a cast of the surface by applying a layer of clear fingernail polish, letting it harden, and then peeling it off. Make a wet mount of the peel or cast; carefully examine under the microscope. On Worksheet 6, precisely draw each specimen—the epidermal cell shapes, stomatal arrangements, and any hairs or glands that are present.

11

WORKSHEET 6. Epidermis of Leafy Vegetables

lettuce	spinach	onion

Q18. If in a criminal case it was important to determine whether a victim's last meal was a lettuce salad with onions from restaurant A or a spinach salad without onions from restaurant B, could examination of the stomach contents answer the question? What is distinctive about each epidermal structure?

☐2. For leaves of *Begonia*, Swedish ivy, *Zebrina*, and three others of your choice, do the following:
• Take an epidermal peel or (if it is not very hairy) make a cast of the surface; mount the peel or cast, and examine.
• Especially for very hairy leaves, cut paper-thin cross sections with a razor blade, slide them on their sides into a drop of water on a slide, cover with a cover slip, and examine.
☐3. On Worksheet 7, carefully draw the epidermis of each specimen; label stomata, subsidiary cells, hairs, and glands.

☐4. Go to the microscope that has the unknown epidermal specimen; determine whether the unknown is the same as one of the leaves that you have examined, or is a totally new one.

Q19. What is the unknown specimen?

_____

III.B. PLASTICITY OF STOMATAL ARRANGEMENT

The basic pattern of stomatal arrangement is genetically determined and can be useful in identifying plant fragments. However, there can be some variation (plasticity) in the basic pattern of stomatal arrangement. In this part of the lab, you will select a hypothesis and investigate it.

MATERIALS

• microscope materials: compound microscope per student, lens paper, microscope slides and cover slips, and dropper bottles of water
• bottles of clear fingernail polish
• plant materials of the students' choice, acquired from the greenhouse or outdoors

WORKSHEET 7. Epidermis of Other Leaves

Begonia	**Swedish ivy**	*Zebrina*
_____	_____	_____

Your question:

Your hypothesis:

Your experimental design: (plants used; location of peels or casts taken; number of replications)

Location of peel	Number of stomata per field of view at _____ power

Your conclusion?

Future investigation needed to clear up any ambiguities?

11

Work in groups.

☐1. Select a question to investigate regarding stomatal arrangement. You can use one of these questions or one of your own:
   a. Is the stomatal density (number of stomata per field of view under a particular objective) the same on sun leaves as on shade leaves of the same plant?
   b. In various plant species with vertically oriented leaves (such as grasses), is the stomatal density equal on the two sides of a leaf?
   c. In various species of plants with horizontally oriented leaves, what is the stomatal density on the upper and on the lower epidermis?
   d. Is the stomatal density the same on a leaf produced early in the growing season and one produced late in the growing season?
   e. Is the stomatal density the same in all parts of one side of one leaf?
   f. Are the stomata of all monocot plants in rows and of all eudicot plants randomly arranged?
☐2. Record your question and hypothesis on Data Sheet 1; then design your experiment.
☐3. Present your experimental design to another group for constructive criticism; then finalize your experimental design and record it on Data Sheet 1.
☐4. Take epidermal peels or fingernail polish casts to investigate your question. Record your data on Data Sheet 1; draw a conclusion.
☐5. Indicate future investigations that would be needed to clear up any ambiguities. Your instructor may require an oral presentation to the class or a written report.

PART IV:
WEIRD LEAVES—MODIFIED FOR SPECIAL FUNCTIONS

OBJECTIVES
1. Become familiar with the diversity of modified leaves.
2. For each modified leaf observed, know its function and its benefit to the plant.
3. Be able to distinguish between a true leaf and a flattened, modified stem.

BACKGROUND

Refer to Figure 11.5 for various modified leaves. Some "leaves" are not even leaves!

• *Phyllanthus* and *Ruscus* have "leaves" that are actually flattened photosynthetic stems. A tip-off is that flowers arise from the edges or the center of these "leaves." True leaves never bear flowers!

Some leaves are just downright spectacular.

• The floating leaves of the giant Victoria water-lilies of the Amazon grow to be over 2.5 meters across, and they are sturdy enough to support a child. Underneath, these leaves are armed with impressive spiny outgrowths. See an image of these leaves on the website of the Kew Royal Botanic Gardens in London at http://www.kew.org/plants/waterlilies/index.html. (If the address has changed, do an Internet search for Kew or an image search for these plants.)

Some leaves are modified for a specific function and may not even look very leaf-like.

• Barberry and cacti have leaves modified into **spines**, as protection against herbivores. (Note that *thorns*

Victoria water lily

Lithops

bladderwort

Figure 11.5 Various kinds of modified leaves

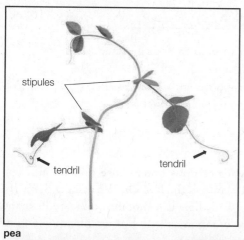

stipules

tendril   tendril

pea

barberry

cactus

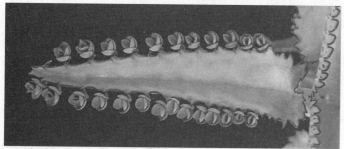

Kalanchoë

3 red bracts

Bougainvillea

Phyllanthus - **leaf-like stems**

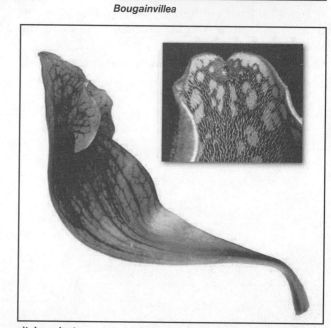

pitcher plant

venus flytrap

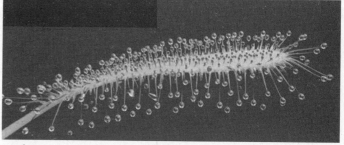

sundew

Figure 11.5 (continued) Various kinds of modified leaves

x

11

y

z

on hawthorn and honey locust are modified branches, while *prickles* on roses are outgrowths of the epidermal tissue.)

- *Bougainvillea*, poinsettia, and others have some of their leaves modified into **bracts**—leaflike, often nongreen structures that often differ in shape from the plant's regular photosynthetic leaves; colorful bracts associated with small inconspicuous flowers can help to attract pollinating animals.
- Pea vines have leaves that consist of several leaflets per leaf; the terminal few leaflets of a pea leaf are modified into **tendrils** that help to support this weak-stemmed plant. (Note that the pea is one of the plants that have two leaflike structures called *stipules* at the base of the leaf stalk.)
- Various insectivorous plants live in nutrient-poor environments; these plants supplement their nutrient intake by capturing and digesting insects with their leaves that are covered with a sticky surface or glandular hairs, or are modified into flytraps, water-filled pitchers, bladder traps, and so forth.
- *Lithops*, *Frithia*, and *Fenestraria* are desert plants that consist of a small cluster of short, thick, flat-topped succulent leaves with translucent "windows" on top to admit light to the photosynthetic cells lining the inside walls.
- *Kalanchoë* produces, on its leaf margin, plantlets that drop off and take root.

MATERIALS

- plant collection in campus greenhouse
- Internet connection
- David Attenborough's *The Private Life of Plants*, Volumes 2 and 6 [videocassettes]

INVESTIGATION

☐1. Tour your campus greenhouse or look at a collection of plants in the lab. On Worksheet 8, record the plants and their leaf modifications; deduce and record the function of each leaf modification and its benefit to the plant.

☐2. If some plants are not available, search the Internet for images, such as by using a Google™ image search.

☐3. To see insectivorous plants close up, view these segments: *The Private Life of Plants*, Volume 2 (an 8-minute segment starting at about 30 minutes into the video) and Volume 6 (a 5-minute segment starting at about 29 minutes into the video, which is then followed by a segment on the giant water-lilies).

WORKSHEET 8. Weird Leaves Observed

Species	Type of modified leaf	Function and benefit to the plant

EXTENDED ASSIGNMENTS

1. Characterize the leaf structure and leaf surface features of a plant of your choice. Draw and label the various slices and epidermal peels that you make.

2. Visit a large botanical garden and/or conservatory near you to view a large diversity of shapes and sizes of leaves. On the website of the American Association of Botanical Gardens and Arboreta (http://www.aabga.org/public_html/index.htm?CFID=1097792&CFTOKEN=76410904), click on "public gardens" and do a search for those in your area. On your next vacation, include a visit to a large botanical garden, such as the Missouri Botanical Garden in St. Louis, Fairchild Tropical Botanic Garden near Miami, Desert Botanical Garden in Phoenix, or Kew Royal Botanic Gardens near London (England).

LITERATURE CITED

American Association of Botanical Gardens and Arboreta. 2006. "Public Gardens Search," http://www.aabga.org/public_html/index.htm?CFID=1097792&CFTOKEN=76410904 (accessed February 19, 2006). Wilmington, DE: American Association of Botanical Gardens and Arboreta.

Attenborough D. 1995. *David Attenborough's The Private Life of Plants*, Volumes 2 and 6 [videocassettes]. BBC/Turner Original Productions, Inc., co-producers. Atlanta, GA: Turner Home Entertainment.

Hangarter, R. 2000. "Plants-In-Motion," http://plantsinmotion.bio.indiana.edu/plantmotion/starthere.html (accessed February 19, 2006). Bloomington: Department of Biology, Indiana University.

Lane, M., L. Anderson, T. Barkley, J. Bock, E. Gifford, D. Hall, D. Norris, T. Rost, and W. Stern. 1990. Forensic Botany: Plants, Perpetrators, Pests, Poisons, and Pot. *BioScience* 40(1):34–39.

Trustees of the Royal Botanic Gardens, Kew. 2006. "Giant Waterlilies," http://www.kew.org/plants/waterlilies/index.html (accessed February 19, 2006). London, UK: Royal Botanic Gardens, Kew.

11

12

LAB

MEIOSIS
CELL DIVISION

I. Meiosis
II. Making offspring
III. The next generation . . . F_2 offspring
IV. Crossing over and its effects

OBJECTIVES

1. Understand the process of meiosis: how a diploid cell with two copies of a particular gene will produce four haploid cells, each having one copy of that gene.
2. Understand the Law of Independent Assortment of two genes located on different chromosomes.
3. Understand the different gametic genotypes that will be produced by a particular diploid parent.
4. Understand the genetic composition of a zygote.
5. Understand how sexual reproduction produces genetic variation, resulting in new genetic combinations (genotypes) and often, in turn, new phenotypes in the population.
6. Understand how crossing over increases even further the genetic variation.

PART I:
MEIOSIS

BACKGROUND

How Does Meiosis Fit into the Plant Life Cycle?

A **diploid** cell has two sets of chromosomes, abbreviated **2n** (**2N** in some books). A **haploid** cell has one set of chromosomes, abbreviated **1n** or simply **n**. Meiosis (Greek: *meioun* = to make smaller) is the special "reduction division" process in the reproductive structures of diploid plants, resulting in the production of haploid **spores** (see Figure 12.1). A haploid spore grows by **mitosis** into a multicellular haploid structure called a **gametophyte**; certain haploid cells of this gametophyte differentiate into haploid **gametes**. These haploid gametes then fuse in the process of **fertilization**, resulting once again in a diploid cell (**zygote**) that then

grows by mitosis into a multicellular diploid plant, the **sporophyte**.

Note that in *animals*, meiosis directly produces gametes—cells capable of engaging in fertilization. There are no spores, and there is no multicellular haploid structure in the animal life cycle.

The benefit of understanding meiosis in detail is that you will then be able to work genetics problems *logically*—you will be able to deduce logically the genetic composition of the gametes that are produced by the parental genotypes in a genetic cross.

Go back and review DNA replication and mitosis *thoroughly* before proceeding with this lab. Fully understanding mitosis will greatly help you in understanding meiosis.

Basic Genetic Terminology and Concepts

As you work on the various parts of this lab, remember the following points:

- A **chromosome** contains a long **DNA** (double helix) molecule.
- A plant cell's genetic information (its **genome**) is encoded in several chromosomes; the exact number of chromosomes is characteristic for each plant species.
- A **gene** is a section of the DNA that encodes a protein that affects a particular trait. For example, there is a *gene* for *flower color* in the garden pea. There are many genes along the length of a chromosome's DNA.
- An **allele** is a particular version of the gene. For example, there is an *allele* for *red flowers* and another *allele* for *white flowers* for the flower *color gene* of the garden pea.
- A **diploid** cell, such as the cells in a fern frond or a flower petal, has two copies of the genome. This means that a diploid cell has two sets of chromosomes, or said another way, it has two copies of each kind of chromosome.

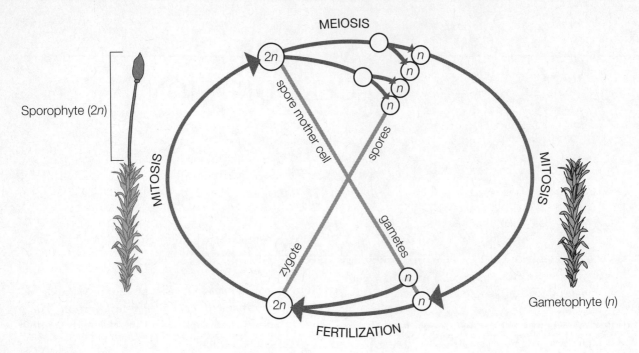

MEIOSIS

Sporophyte (2n)

MITOSIS

spore mother cell

spores

MITOSIS

zygote

gametes

Gametophyte (n)

FERTILIZATION

Figure 12.1 The plant life cycle (moss as the example)

- A **haploid** cell has half the diploid chromosome number; a haploid cell has one genome. This means that there is one set of chromosomes; there is one copy of each kind of chromosome.
- In a diploid cell, two chromosomes of the same kind (one from one chromosome set and one from the other set) have the same array of genes along their length, and are called **homologous** chromosomes. Two homologous chromosomes (**homologues**, also spelled **homologs**) may bear different alleles for a particular gene position; the array of genes along homologous chromosomes is identical, but not necessarily the alleles at those gene positions.
- The **genotype** of a cell is the genetic makeup of the cell. By convention, the same letter of the alphabet is used as a symbol for the alleles of one gene. Here is the key to the genes and allele symbols on your chromosome models:

Gene for flower color:	Gene for plant height:	Gene for branching:
R = red r = white	T = tall t = short	B = branched b = single stem
Gene for leaf shape:	**Gene for leaf arrangement:**	**Gene for fragrance:**
E = elongated e = rounded	A = rosette a = along stem	F = fragrant f = not fragrant

- If you are examining only the gene for flower color and the gene for branching, then the genotype for

your model chromosomes is *RrBb*. If you are examining two genes that are **linked** (on the same chromosome) such as the flower color and the height genes on your model chromosomes, then the genotype of your model, *RrTt*, can also be written *Rt/rT* to show that alleles *Rt* are together on one homologue and alleles *rT* are on the other homologue.

- The **phenotype** is the physical manifestation of the genotype. For the flower-color gene, allele *R* is for red flowers and allele *r* is for white flowers. If a plant has one of each allele (genotype *Rr*) and has red flowers, then the "red" allele that determined the phenotype is called **dominant**. The "white" allele, whose effect is not visible, is called **recessive**. By convention, the dominant allele is symbolized by the *capital* letter, and the recessive allele by the *lowercase* letter. In the accompanying "key to the genes", the capital letter indicates the dominant allele for each gene.
- A plant that is genotype *Rr* is said to be **heterozygous** for this gene. The genotype *RR* is called **homozygous dominant**, while *rr* is **homozygous recessive**.
- A diploid cell undergoing **meiosis** actually undergoes two divisions back-to-back, **meiosis I** and **meiosis II**, that result in the production of four haploid cells.

OK, now that you have the basics down, let's do meiosis! If your lab table is too small to accommodate all the felt sheets of your group, then you can either hang your felt sheet on the wall or work on the floor.

MATERIALS

- red, navy, pink, and light-blue poster boards; scissors, felt-tip pens, six colors of non-black sticky-backed labels, sticky-backed Velcro®, chalk, masking tape, and transparent tape
- the chromosome models that you made out of poster board and Velcro® as instructed in the mitosis lab earlier in the course, along with your felt sheet that represents the cell.
- In your group of four students, one student made red chromosomes, another made navy blue ones, another made pink ones, and another made light-blue ones.
- Each chromosome model set has three pairs of homologous chromosomes—a long pair of genotype *RrTt*, a medium pair of genotype *BbEe*, and a short pair of either genotype *AAFF* for the red and the pink chromosome models, or genotype *aaff* for the navy and the light-blue chromosome models.

INVESTIGATION

☐1. *Using your chromosome models, represent a* <u>*diploid*</u> *cell with* <u>*unreplicated*</u> *chromosomes by attaching the appropriate labeled chromosomes to your felt sheet.* There should be two homologous long, two homologous medium, and two homologous short chromosomes; two homologues have the same linear array of genes (letters) although the alleles may be different (capital or lowercase, signifying dominant or recessive).

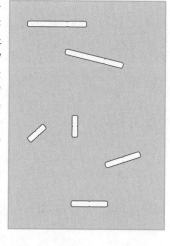

☐2. *Replicate the chromosomes* in preparation for **meiosis I**, as you did for mitosis. Check each replicated chromosome to make sure that the two replicas are <u>*identical*</u> in their labeled **alleles**. Replicated chromosomes condense (become very compact) during *prophase I* of meiosis I, as was the case in prophase of mitosis.

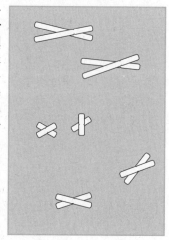

☐3. *Now demonstrate* *synapsis*—which appears *only in prophase I of meiosis I and lasts through metaphase I*—by positioning the two replicated long chromosomes together, the two replicated medium chromosomes together, and the two replicated short chromosomes together.

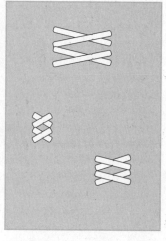

☐4. Also in *prophase I*, **spindle fibers** from one pole attach to the centromere of one replicated chromosome of the synapsed pair, and spindle fibers from the *other* pole attach to the centromere of the *other* replicated chromosome of the synapsed pair. During spindle fiber attachment, the synapsed chromosomes become positioned at the spindle's equatorial plane.

☐5. *Now demonstrate* **metaphase I** — the stage at which synapsed chromosomes are at the spindle's equatorial plane—by drawing chalk lines to represent fibers and positioning your chromosomes.

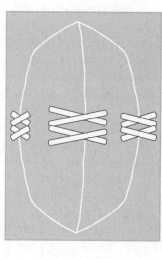

Note that replicated chromosomes are arranged differently in meiosis metaphase I than in mitosis metaphase!

☐6. *NOTE*: The medium chromosomes can be lined up in either of two ways relative to the long chromosomes: with the medium chromosome bearing allele *B* being on the *same* side or on the *other* side of the equator as the long chromosome bearing allele *R*. It is a 50–50 random chance as to how the second pair lines up relative to the first pair of chromosomes. *In lab, flip a coin whenever there is a choice to be made as to the arrangement of a synapsed pair on the spindle's equator.* This randomness of orientation of a synapsed pair at the equator is the basis of the **Law of Independent Assortment:** Genes on one kind of chromosome assort into the two new cells independently of the genes on another kind of chromosome.

12

□7. *Now demonstrate* **anaphase I** *— the spindle fibers radically shorten and one replicated chromosome of the synapsed pair moves toward one pole and the other replicated homologue moves toward the other pole.*

NOTE: When the replicated chromosomes are *synapsed* in metaphase I, subsequent anaphase movements result in *reduction division (meiosis)* rather than mitosis. See Figure 12.2, which compares mitosis and meiosis.

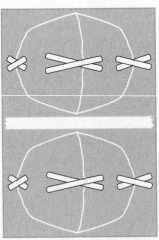

□8. *Continue moving the chromosomes until they are clustered at the poles*, which defines **telophase I**. In late telophase I, the cytoplasm divides between the two clusters of chromosomes to form two new cells, completing meiosis I. *Indicate the two new cells by running a strip of masking tape down the middle of the felt sheet.*

□9. The two new cells, however, are not yet at the point of having unreplicated chromosomes; to accomplish this, a second division occurs immediately. For meiosis II, use the back side of the felt sheet, which is not yet marked up with chalk "spindle fibers"; *transfer the chromosomal content of the two cells from meiosis I along with the masking tape strip onto the back side of the felt.*

□10. During **prophase II** of the second division, **meiosis II**, new spindle fibers from both poles attach to the centromere of each still-replicated chromosome. *Draw chalk lines and position the chromosomes on the equator to demonstrate* **metaphase II**.

□11. *Demonstrate* **anaphase II** in each cell by pulling the Velcro of the two replicas of each replicated chromosome apart and moving the two replicas, now finally called **chromosomes**, to opposite poles.

□12. When the chromosomes reach the poles, the cells are in **telophase II**. Cytoplasmic division follows. *Represent this by putting masking tape across each cell.* Thus the two cells from meiosis I are now four cells after meiosis II.

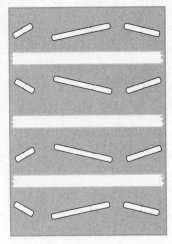

□13. Check your work! Does each of your four cells after meiosis have only one unreplicated chromosome of each kind—one long, one medium, one short? If so, then you have succeeded in making four haploid cells.

□14. Write the genotypes of each of the four haploid cells that you produced:

_____

_____

_____

_____

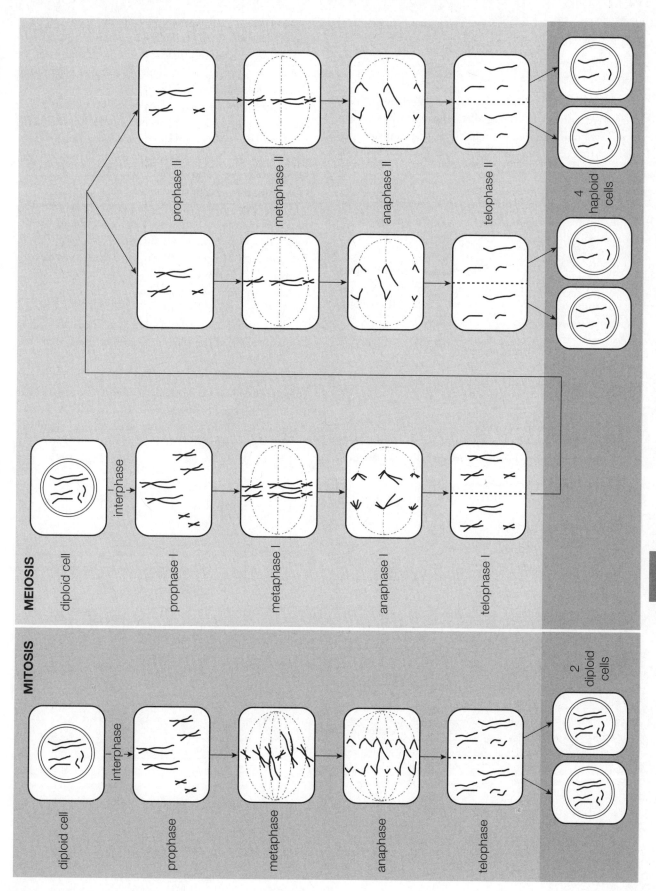

Figure 12.2 Comparison of mitosis and meiosis

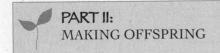

PART II:
MAKING OFFSPRING

In plants, the four haploid cells produced by meiosis are called **spores**; a spore will grow via **mitosis** into a multicellular haploid structure, having certain cells that mature into haploid **gametes**. Because mitosis "makes more of the same," a gamete will have exactly the same chromosomal content as that haploid spore had. Thus the chromosomes in one of the cells that you produced by meiosis in Part I can represent the chromosomes in the resulting gamete.

In this part of the lab, you will investigate the offspring produced by **sexual reproduction**, which involves the fertilization of an egg by a sperm. You will combine the chromosomes of one of the cells you produced in Part I with those of one of the cells produced by another student.

INVESTIGATION

☐1. Randomly select one of the four haploid cells that you have produced via meiosis in Part I to represent a gamete. Pick up the chromosomes of your selected cell.

☐2. *Students with the red or the pink models*: Red and pink models represent the female parents, and their meiosis process ultimately leads to the production of an egg cell—represented by the cell that you selected in the previous step.

☐3. *Students with the navy or the light-blue models*: Navy and light-blue models represent male parents, and the meiosis process ultimately leads to the production of sperm—represented by the cell that you selected in step 1 above. You will donate this cell's chromosomes to a particular female's egg.

☐4. *Cross the red parent with the navy-blue parent*: The haploid egg (red chromosomes) is fertilized by the haploid sperm (navy chromosomes), resulting in the pooling of the chromosomes of these two cells to produce the diploid **offspring**.

☐5. *Cross the pink parent with the light-blue parent*: The haploid egg (pink chromosomes) is fertilized by the haploid sperm (light-blue chromosomes), resulting in the pooling of the chromosomes of these two cells to produce the diploid offspring.

NOTE: If there is not an equal number of male and female chromosome models in the class, a male can provide sperm chromosomes to more than one female, or a female could have several eggs fertilized by different males.

DATA SHEET 1: Offspring Plant

Female Parent		
Genotype	**Phenotype**	**Appearance**
Rr *Tt* *Bb* *Ee* *AA* *FF*	red flowers tall branched elongated leaf rosette fragrant	

Male Parent		
Genotype	**Phenotype**	**Appearance**
Rr *Tt* *Bb* *Ee* *aa* *ff*	red flowers tall branched elongated leaf leaves on stem not fragrant	

Your Offspring Plant		
Genotype	**Phenotype**	**Appearance**

☐6. Put the offspring's chromosomes onto a felt sheet from which all remaining chromosomes have been cleared. In the diploid offspring, each homologous pair of chromosomes consists of one chromosome from the mother (red or pink) and one chromosome from the father (navy or light blue).

Q1. In the diploid offspring, did the female parent and the male parent contribute the same number of chromosomes? _____ Did each contribute one of each kind of chromosome? ____

☐7. Determine the genotype and phenotype of your offspring, and fill in Data Sheet 1.

Q2. Did *your* offspring plant have a different phenotype than either parent? _____

☐8. Also record your results on the classroom board, in a chart with column headings "genotype" and "phenotype"; if your offspring's phenotype was *different* from that of both parents, then put a star by it.

☐9. From the compiled class data, answer these questions:

Q3. How many total offspring were "produced" by your class? _____

Q4. What *fraction* of the total offspring "produced" by the class consisted of phenotypically new individuals (*different* from both parental phenotypes)? _____

Q5. *Among the new phenotypes generated* in the class via sexual reproduction, how many *different kinds* of new phenotypes were there? _____

Q6. What two features about sexual reproduction and meiosis resulted in the greater diversity of phenotypes in the offspring?

PART III:
THE NEXT GENERATION . . .
F$_2$ OFFSPRING

For ease in communication, the original parents are called the parental (**P**) generation; their offspring which you produced in Part II are called the first filial (**F$_1$**) generation, while the offspring's offspring are called the second filial (**F$_2$**) generation.

Let's determine the type of F$_2$ offspring produced by crossing the F$_1$ offspring from the red and navy parental mating with the F$_1$ offspring from the pink and light-blue parental mating. This requires each F$_1$ offspring to go through meiosis to produce cells representing gametes; then fertilization can occur to produce the F$_2$ offspring.

INVESTIGATION

☐1. Working with your lab partner, place your "F$_1$ plant" chromosomes from Part II of this lab onto the felt sheet. At each lab table of four students, there should be one pair of students whose F$_1$ offspring plant resulted from a red with navy parental mating, and the other pair of students whose F$_1$ offspring plant resulted from a pink with light-blue parental mating. *Record* the genotype and phenotype, and *draw* the appearance of your F$_1$ plants in Data Sheet 2.

☐2. To **replicate** each of the chromosomes of your F$_1$ plant, cut a replica from the *matching color* of poster board, stick on labels, and write allele symbols on the labels to give an exact replica; create a Velcro® centromere with the *fuzzy* patch on the labeled side of the replica. Attach your replica to the chromosome. Check to make sure that the two replicas of a chromosome are identical—the same color and same sequence of alleles.

☐3. Move your chromosomes into **synapsis**.

☐4. Now show metaphase I. Be sure to *flip a coin* to randomly determine which replicated homologue lies on which side of the spindle's equator. Now proceed to anaphase I and telophase I to complete meiosis I.

☐5. Follow with the steps of meiosis II, concluding with cell division.

☐6. The four cells that you produced by this meiosis will give rise to gametes. Randomly select one of your four cells to represent a gamete. Demonstrate fertilization: if you are working with a red x navy F$_1$ plant, then combine your selected gamete with that of a pink x light-blue F$_1$ plant, and vice versa. Place

DATA SHEET 2: Production of F₂ Plant

Red x Navy F₁ Plant		
Genotype	Phenotype	Appearance

Pink x Light-blue F₁ Plant		
Genotype	Phenotype	Appearance

Your F₂ Plant		
Genotype	Phenotype	Appearance

the chromosomes of each gamete together onto one felt sheet.

☐7. You have just produced a zygote that via mitosis will grow into an F₂ plant. In Data Sheet 2, record the genotype and phenotype, and draw the appearance of your F₂ plant.

☐8. Record on the classroom board your F₂ plant's genotype and phenotype, and put a star by it if it has a novel phenotype never before seen in the P and the F₁ generations in class.

Q7. Did each grandparent contribute equally to the F₂ plant? Record the chromosomes that were inherited from the grandparent that was:

red_____

navy_____

pink_____

light blue_____

Q8. Did your F₂ plant produced by sexual reproduction have a new phenotype that was shown by neither of its two parents (the F₁ plants) nor its four grandparents (the P-generation plants)?

EXTENDED ASSIGNMENT (thought questions)

When you look through seed catalogs, you occasionally see plants listed as being **triploid** and heavy blooming. Triploids are generally not capable of producing seeds (with three copies of each kind of chromosome, the pairing of chromosomes in synapsis and meiosis I is not normal). Thus, energy that would normally go into seed production will instead go into flower production.

If two gametes underwent fertilization and produced a triploid zygote, what ploidy would each gamete have had?

Each of these gametes had been produced by reduction division (meiosis); therefore, what was the ploidy of each of the two parents that were crossed to produce this triploid?

If a large number of triploid plants appeared in a natural ecosystem, what effect would that have on the ecosystem over the years? Consider the vegetation as well as the animals that visit plants for various reasons. Consider the effect if these triploids were all annuals (plants that live one year) versus the effect if these were all perennials (live more than two years).

PART IV:
CROSSING OVER AND ITS EFFECTS

NOTE: Do this section *only after* you *thoroughly* understand meiosis.

You have seen that the alleles for stem branching (*B* and *b*) on the medium chromosome *assorted independently* from the alleles for flower color (*R* and *r*) on the long chromosome, due to the *random orientation* of synapsed chromosomes on the equator at metaphase I. The *B* allele for branching was equally likely to end up in a cell with allele *R* as with allele *r* of the *un*linked flower-color gene.

However, the stem-branching gene and the leaf-shape gene are *linked*, both on the medium chromosome. So far during meiosis, we have seen the *B* and *E* alleles of these genes always traveling together because they were on the same homologue, and likewise the *b* and *e* alleles traveled together because they are on the other homologue of the medium chromosome.

Let's produce the F$_2$ again, but see what happens when you add **crossing over**—a routine process that occurs during synapsis when replicated homologous chromosomes lie alongside each other. Crossing over occurs when a replica from one replicated chromosome and a replica from the *homologous* replicated chromosome break at the same place, and the pieces rejoin to the wrong chromosome (see Figure 12.3). The two replicas of a replicated chromosome are called **sister chromatids** as long as they are joined at the centromere; once separated, they are called chromosomes. Two chromatids that are part of different replicat-

ed homologues are called **non-sister chromatids**. Thus, crossing over is also referred to as **non-sister chromatid exchange**.

INVESTIGATION

☐1. Repeat the procedure of Part III, steps 1–3.

☐2. When the chromosomes are synapsed, do crossing over between two *non-sister chromatids* of *homologous* chromosomes by cutting the poster-board chromatids at the *same* location between two genes, switching the cut pieces, and taping each piece onto the cut end of the non-sister chromatid (use transparent tape). If you want to do more than one crossing-over event in your cell, go ahead!

☐3. Continue with steps 4–8 of Part III; record data on Data Sheet 3.

Q9.   What genetic material did your F$_2$ plant receive from each of the grandparent plants:

red_____

navy_____

pink_____

light blue_____

Q10. Did you or others in the class get an F$_2$ phenotype that would have been impossible to get without crossing over? Describe and explain.

12

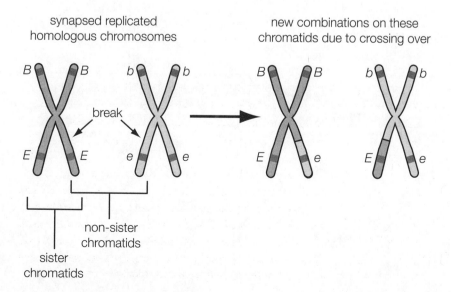

synapsed replicated homologous chromosomes

new combinations on these chromatids due to crossing over

Figure 12.3 Crossing over

Q11. Discuss the features of sexual reproduction that promote genetic variation in a population.

Q12. Discuss how genetic variation promotes long-term survival of a species.

DATA SHEET 3: Crossing Over During Production of F$_2$ Plant

Red x Navy F$_1$ Plant		
Genotype	**Phenotype**	**Appearance**

Pink x Light-blue F$_1$ Plant		
Genotype	**Phenotype**	**Appearance**

Your F$_2$ Plant		
Genotype	**Phenotype**	**Appearance**

13

LAB

PLANT LIFE CYCLES

I.   Culturing moss and fern spores
II. C-FERN® life cycle: chemotaxis of sperm

NOTE: The objectives are given at the beginning of each part of this lab.

BACKGROUND

All plants—seedless plants as well as seed plants—have the type of life cycle called *alternation of generations*, which is illustrated in Figure 13.1. Investigations in this lab focus on the life cycle of the seedless plants because both of their generations are large enough to be easily seen. Note that the diverse group called the algae is not considered to be part of the plant kingdom.

Life Cycle of Seedless Plants

In the life cycle of all **seedless plants** (mosses, ferns, and their close relatives), <u>spores are the structures for dispersal to a new location</u>. A spore is a *single cell*. That spore will germinate and grow into a multicellular structure called a **gametophyte**. Gametophytes produce **gametes**; the gametes of seedless plants consist of nonmotile eggs and *flagellated* sperm cells. Gametes are single cells that can engage in fertilization. The flagellated sperm swims through external water to the egg and fuses with it in fertilization. The fertilized egg (called a **zygote**) remains in place and grows into the multicellular structure called a **sporophyte**. The spo-

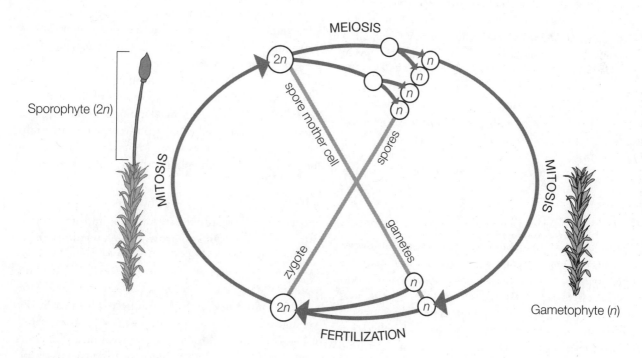

Figure 13.1 The plant life cycle (moss as the example). Note that every time you cross a line of the "X" in the diagram, there is a special cell that then undergoes a particular process to form something else.

NOTE: Your instructor may have the class do only certain parts of this lab or may have different student groups do different parts and then report their results to the class.

141

rophyte produces **spores**, which are then dispersed, completing the life cycle. In the plant life cycle, the multicellular gametophyte alternates with the multicellular sporophyte; such a life cycle is thus called **alternation of generations**.

- *Mosses.* The leafy moss plant is the gametophyte; at the top of this plant, the gametes (egg and sperm cells) are produced and fertilization occurs. The fertilized egg (zygote) stays in place and grows into a sporophyte consisting of a stalk topped by a capsule, inside of which the spores are produced.

- *Ferns.* The fern's gametophyte is easily overlooked; it is a thin, heart-shaped structure about the size of your little fingernail or smaller. This gametophyte produces the gametes; the fertilized egg (zygote) stays in place and grows into the large fern frond. On the underside of the fern frond, the spores are produced.

Life Cycle of Seed Plants

Seed plants include the gymnosperms (conifers and others) and the angiosperms (flowering plants). The seed plants likewise have the alternation of generations life cycle. The seed plant's spores, however, remain in place and develop into microscopic gametophytes consisting of only a few cells. The male gametophyte is the **pollen grain**, which contains the sperm. The pollen grain is transported by wind, water, or animal pollinator to the vicinity of the nonmotile egg cell; then the pollen grain grows a pollen tube through the tissue surrounding the egg to deliver the sperm cell to the egg for fertilization. The tiny embryonic sporophyte, along with stored nutrients, is enclosed in the protective seed coat of a seed—the structure for dispersal to a new location. (See your textbook and Lab 6, "Seed Germination", Lab 14, "Patterns of Inheritance", and Lab 18, "Reproduction of Flowering Plants," for details.)

The Genetics of the Plant Life Cycle

A spore contains one complete set of genetic information (has one copy of each kind of chromosome) and is called **haploid**, abbreviated $1n$ or n (capital letter N is used in some books). By mitosis cell division, the haploid spore grows into the haploid gametophyte, which produces haploid sperm and egg cells. Fertilization fuses a sperm and egg to form the zygote that is diploid (2n)—has two complete sets of genetic information. By mitosis cell division, the diploid zygote grows into the diploid sporophyte, which by means of reduction division, meiosis, produces haploid spores.

PART I: CULTURING MOSS AND FERN SPORES

OBJECTIVES

1. Practice the scientific method and proper experimental design in a long-term study.
2. Develop an optimal culturing procedure for spores of mosses and of ferns.
3. Observe the stages of, and the developmental differences in, the life cycle of mosses and ferns.

BACKGROUND

"Plant Food"

Plants require carbon dioxide, water, and solar energy for photosynthesis to make energy-rich sugars. In addition to the carbon, hydrogen, and oxygen derived from carbon dioxide and water, plants also need many other elements (often called minerals) for the chemical processes and structures in living cells. For example, chlorophyll also contains nitrogen and magnesium, proteins contain nitrogen and sulfur, and DNA (genetic material) contains nitrogen and phosphorus. Elements such as nitrogen (N), phosphorus (P), potassium (K), magnesium (Mg), and sulfur (S) are examples of "macronutrients" since they are needed in relatively large amounts by plants. Other elements such as iron (Fe), copper (Cu), manganese (Mn), and zinc (Zn) are examples of "micronutrients" since they are required in relatively small amounts. What people call "plant food" or "fertilizer" is a mixture of the elements that tend to be depleted most quickly from the soil. There is still much to be learned about plant nutrition; for example, boron is a known micronutrient but the details of its role in the plant cell are still unknown.

Questions

Spores, sprinkled onto the surface of liquid culture medium, will germinate and grow. Can you culture moss and fern spores on diluted all-purpose plant food, or do you need to buy special culture medium? If plant food is adequate for houseplants, might it work for spore germination? Will the dilution specified on the package work, or does it need to be more concentrated or less concentrated for spore germination? Will plain water work? Is there any difference in the needs of the moss and the fern spores? Let's investigate this potential cost-saving method of culturing spores by trying a wide range of plant-food dilutions.

MATERIALS

- moss spores (Carolina RG-15-6661)
- fern spores (Carolina RG-15-6860)
- package of all-purpose plant food
- distilled water
- culture dishes
- balances
- plastic wrap
- rubber bands
- tape
- felt-tip permanent markers
- graduated cylinders, various sizes
- liter beakers, for mixing
- living moss plant with sporophyte (Carolina ER-15-6695)
- living fern plant with sori (Carolina ER-15-6902)
- dissecting microscopes

INVESTIGATIONS

☐1. Get started by first becoming familiar with your plant food: Look at the plant-food package's front label; the three numbers (such as 15-30-15 or 20-20-20) are the **N-P-K** numbers indicating the % nitrogen, % phosphorus, and % potassium for the mix. These three macronutrients are the ones most likely to become depleted in soils. Different plants prefer different ratios; thus, devoted gardeners buy a special formulation for tomatoes to ensure a good fruit set, another for acid-loving plants, another for African violet, and so on.

Q1.   What are the N-P-K numbers of your plant food?

☐2. Now look at the *list of ingredients* or *"guaranteed analysis"* in small print elsewhere on the package; note that in addition to N, P, and K, the plant-food mix typically also contains half a dozen or so of the micronutrients. Notice that they are present at much lower levels in the mix.

Q2.   What micronutrients are also in your plant food?

☐3. On the plant food's label, there are instructions for diluting the granules to a concentration that is appropriate for periodically watering the average houseplants or vegetables. Often a plastic measur-ing scoop is included. You could use the scoop or you could weigh 1 measure of granules and subsequently always use *weight*.

Q3.   Which method is more precise: scoop or weight?

☐4. Examine the living moss and fern plants, if available, to see where the spores are produced. Spores on a live plant might not be mature yet, or they might already have been released from the plant. Thus, vials of viable spores are provided for you to use in your investigation. If by chance your plants are in an active spore-shedding stage, you could try using the spores directly from the plant.

☐5. The actual experiment will be performed by the entire class working together. Work in groups of 4–6 students to discuss, brainstorm, and develop an experimental procedure for the class that will test a wide range of plant food dilutions to determine the best culture solution for moss and for fern spores. *Remember the hallmarks of the scientific method: replication, control, and one variable only.* Determine the dilutions to be used, and the procedure to be followed in making these dilutions. Record on Worksheet I.

☐6. Discuss as a class the proposals of the groups, and develop the best experimental design for the class. Cover a wide range of dilutions, and include replication. Also decide how to parcel out the work among all of the class members.

☐7. Make the dilutions with distilled water, fill the dishes to the level decided upon, mark the level of liquid with permanent marker, and then lightly sprinkle a few spores onto the surface (by lightly tapping the vial's spore-covered cotton plug above the dish). If you look against the light, you will just barely see the spores on the liquid surface. Cover the dish with plastic wrap held in place by a rubber band; place the covered dishes near a window or under lights.

☐8. On Worksheet I, prepare a table in which you will record your weekly observations.

☐9. Each week, use a dissecting microscope to check the surface of the liquid for spore germination and gametophyte growth. Record observations; include a drawing of the developing gametophyte each week. Remember that when a culture dish is partially full of liquid and covered with plastic wrap held on by a rubber band, there might be gaps that permit evaporation. Thus, the level of the liquid in the culture dishes needs to be marked at the start, then checked and maintained each week.

13

WORKSHEET I: Culturing Moss and Fern Spores

Experimental Design:

Data:

Q4. Why should you mark the level of the liquid in the dishes and maintain that level?

☐10. If after two weeks there still is no evidence of germination and growth, then the plant-food concentrations may be either too strong or too weak. Discuss with your classmates, and revise the experimental design. Set up another set of culture dishes with the new dilutions.

☐11. If growth is visible only in the dish with the weakest or the strongest plant-food dilution, prepare a new set of dishes to test a variety of concentrations around that level in order to determine the optimum condition more accurately.

☐12. Observe your culture dishes for the rest of the term, comparing and contrasting the development of the moss and the fern gametophytes under the different nutrient conditions.

Q5. What is the optimum dilution for *moss* spores? _____ g/100ml

For *fern* spores? _____ g/100ml

Q6. What do your results suggest about the degree of competition between moss and fern gametophytes on different substrates in nature?

Q7. Do you even need nutrients, or did plain water work for culturing?

Q8. Compare and contrast the structure of the developing gametophytes of moss and of fern.

PART II:
C-FERN® LIFE CYCLE: CHEMOTAXIS OF SPERM

II.A. FERN LIFE CYCLE

OBJECTIVES

1. Observe and understand the rarely observed stages of the fern life cycle: the spores, gametophytes, motile sperm, and origins of sporophytes.
2. Understand the environmental factor that triggers sperm release.

MATERIALS

- Sex in a Dish: C-FERN® Life Cycle Kit (Carolina ER-15-6702); this kit provides all the materials, background information, drawings, and instructions for observing the life cycle of *Ceratopteris*, a tropical fern that likes warm and humid environments.
- humid environment with continuous bright light and approximately 28°C for growth of the cultures in the culture dome that is included in the kit. If no such incubator is available, then use:
  - C-FERN Growth Pod™ (Carolina ER-15-6715)
  - lighting fixture with 6-inch dome and switch
  - 15W fluorescent bulb
  - clamp
  - support stand
  - small thermometer to put in the growth pod
- dissecting microscopes

INVESTIGATIONS

☐*With this kit:* You will follow the kit's instructions and make weekly observations:

Week 1: You sow the spores onto solid medium in a petri dish.

Week 2: You examine the germinating spores under a microscope.

Week 3: You observe the release of sperm in response to the addition of water, and you observe the swimming of the sperm to the egg cells.

Week 4: You observe the early development of the sporophyte.

Until the end of the term: You make weekly observations of the developing sporophyte.

Follow the instructions supplied with the kit, and enjoy!

13

Q9. How is reproductive success promoted by sperm release being triggered by the addition of water rather than by some other factor?

Q11. How does chemotaxis of sperm promote reproductive success in a natural ecosystem?

II.B. SPERM CHEMOTAXIS

OBJECTIVES

1. Observe the specificity of chemotaxis of fern sperm.
2. Understand the need for chemotaxis in nature.

Q12. All of the seedless plants produce flagellated sperm. In what kinds of terrestrial ecosystems would you expect the seedless plants to be most abundant? Least abundant? Explain.

MATERIALS

- Chemical attraction: C-FERN® Sperm Chemotaxis Kit (Carolina ER-15-6714) supplies all the materials and instructions for observing the chemotaxis of sperm cells.
- 12- to 18-day-old cultures of C-FERN® gametophytes, which you have grown with the previously mentioned C-FERN® Life Cycle Kit (Carolina ER-15-6702)
- dissecting microscopes with external light sources

Q13. Review the information on the life cycles of seedless plants and seed plants at the beginning of this lab. Seedless plants move to a new location as spores; seed plants do produce spores, but seed plants move to a new location as seeds. What advantages does the seed plant life cycle have over the seedless plant life cycle?

INVESTIGATIONS

☐ *With this kit:* You will follow the kit's instructions and determine the degree of response of swimming sperm to succinic acid, L-malic acid, D-malic acid, fumaric acid, and maleic acid.

NOTE: When looking for the sperm under the microscope, angle the light until the background is black; then the sperm will look like tiny, swarming specks of light.

Q10. Would you expect the motile sperm of ferns to have a chemotactic response? Why?

FURTHER INVESTIGATION

Determine the chemotactic response of C-FERN® sperm to several other materials. For example, test the effect of salinity (salt solution) or acidity (vinegar). If you see a chemotactic response, positive or negative, consider the effect that such pollutants in the environment could have on the reproductive success of these ferns and perhaps other seedless plants.

LAB 14

PATTERNS OF INHERITANCE

I. Inheritance of one gene
II. Inheritance of two genes
III. Corn genetics: evaluating observed vs. expected
IV. Further investigations

NOTE: The investigations of Part III of this lab require an understanding of Parts I and II. Thus, Parts I and II may be given as pre-lab assignments to be discussed in lab before the investigations of Part III are conducted.

OBJECTIVES

1. Given the genotype of an individual, be able to give the genotypes of its gametes and the genotypic frequencies (Parts I and II).
2. Given the parental genotypes for one trait, be able to predict the expected genotypic ratio and phenotypic ratio in the offspring generation (Part I).
3. Given the parental genotypes for two traits, be able to predict the expected genotypic ratio and phenotypic ratio in the offspring generation (Part II).
4. Understand sexual reproduction in corn, including the genetic composition of the various parts of a corn kernel (Part III).
5. Understand that several genes may be required to produce a particular trait (Part III.C).
6. Be able to use the chi-square test to determine whether the observed genotypic frequency in the offspring generation agrees with the calculated expected genotypic frequency (Parts III.A through III.D).
7. Understand that outcrossing sexual reproduction produces greater phenotypic diversity than does selfing sexual reproduction, and that selfing produces greater phenotypic diversity than does asexual reproduction (Part IV.A).
8. Understand the effect of gene dosage of certain genes on the phenotype of the endosperm (Part IV.B).

BACKGROUND

For a more thorough understanding of the material covered in this lab, review the topics of meiosis, flower structure, and sexual reproduction in plants before doing this lab. Key concepts are presented here:

- A genetic trait such as flower color or seed shape is the result of genetic information that is in the chromosomes, which were inherited from the parents via the sperm and egg.
- In plants and other organisms that are not bacteria, one complete set of genetic information consists of several linear chromosomes.
- A **chromosome** contains a long **DNA** (double-helix-shaped) molecule.
- A **gene** is a section of the DNA that encodes a protein that affects a particular trait. For example, there is a *gene* for *flower color* in the garden pea. Many genes exist along a chromosome's DNA.
- An **allele** is a particular version of the gene. For example, there is an *allele* for *red flowers* and another *allele* for *white flowers* for the flower-color gene of the garden pea.
- A gamete (sperm or egg cell) is called **haploid** (abbreviated $1n$) because it carries *one* set of chromosomes—and, consequently, one of each kind of gene, typically.
- Fertilization is the fusion of the sperm with the egg, which combines their chromosomal contents. The resulting zygote contains two sets of chromosomes and is thus called **diploid** (abbreviated $2n$). The diploid zygote grows into a diploid adult body.
- A diploid cell has *two* of each kind of gene, typically, because it has two of each kind of chromosome. For a particular gene in a diploid cell, the two *alleles* present (one on each chromosome) may or may not be the same kind of allele.

14

NOTE: Your instructor will indicate the parts of this lab that you will be doing before class, during class, and possibly as homework or extra credit after class.

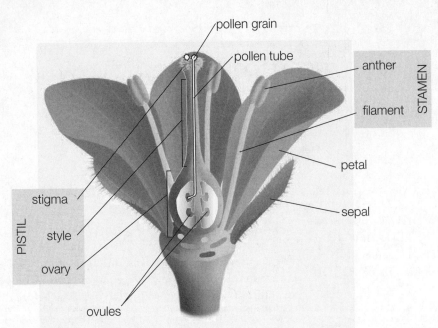

Figure 14.1 Flower with a pollen tube that has grown into an ovule

pollen grain

pollen tube

anther

filament

STAMEN

petal

stigma

PISTIL

style

sepal

ovary

ovules

- The **genotype** of a cell is the genetic makeup of the cell. By convention, the *same letter of the alphabet* is used as a symbol for the alleles of one gene. Thus, for the flower-color gene of the garden pea, allele *R* is for red flowers and allele *r* is for white flowers.
- The **phenotype** is the physical manifestation of the genotype. If a plant has one of each allele for the flower-color gene (genotype *Rr*) and has red flowers, then the allele for redness determined the phenotype and is called **dominant**. The allele for whiteness does not affect the phenotype of this plant, and is called **recessive**. (For the recessive phenotype to occur, two copies of the recessive allele are needed—genotype *rr*.) By convention, the dominant allele is symbolized by the capital letter, and the recessive allele by the lowercase letter.
- A plant that is genotype *Rr* for flower color is said to be **heterozygous** for this gene. The genotype *RR* is called **homozygous dominant**, while *rr* is **homozygous recessive**.
- A heterozygous organism (*Rr*) produces two kinds of gametes and they are in equal proportions: Half of the gametes contain allele *R* and half contain allele *r*. (This is known as Gregor Mendel's First Law of Heredity or the Law of Segregation.)

In flowering plants, two sperm cells are contained within each pollen grain produced within the anther. One egg cell is contained within each ovule that is in the ovary of the pistil. Wind or various animals typically transport the pollen onto the stigma at the end of the pistil; this is called **pollination**. The receptive stigma secretes a sugary liquid, in which a pollen grain germinates and then grows a pollen tube down through the style into the ovary and into one ovule (see Figure 14.1). A sperm cell traveling down the pollen tube fuses with or **fertilizes** the ovule's egg cell. The mechanism of pollen tube guidance from the stigma to the egg is an active area of research.

To "perform a genetic cross" with flowering plants means that the researcher takes pollen from the paternal parent and deposits it onto the stigma of a pistil in a flower on the maternal parent. If a **selfing** cross is performed, then the pollen and stigma involved in the pollination are of the *same* plant. If **outcrossing** is to be performed, selfing must first be prevented by removing the flower's own anthers before they can open and shed their pollen; then pollen from a *different plant* is used to pollinate that flower. After the flower is hand pollinated, it can be covered with a small bag to prevent any subsequent natural pollination with pollen from unknown sources.

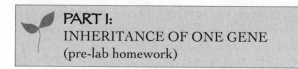

Example 1

Gene for flower color has alleles:
 R = red-flowering allele, dominant
 r = white-flowering allele, recessive

Cross a heterozygous red-flowering pea with a white-flowering pea. In other words, take pollen from one of these plants and use it to pollinate the other plant.

In symbols:

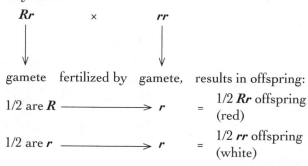

Genotypic ratio expected in the offspring generation is 1 Rr : 1 rr (1/2 Rr, 1/2 rr).

Phenotypic ratio expected in the offspring generation is 1 red : 1 white (1/2 red : 1/2 white).

This process of working the problem could be abbreviated by writing the information in a Punnett square:

		possible genotypes of gametes from other parent
		all are r
possible genotypes of gametes from one parent	1/2 R	1/2 Rr offspring
	1/2 r	1/2 rr offspring

Example 2

Gene for flower color has alleles:
 R = red-flowering allele, dominant
 r = white-flowering allele, recessive

Self-pollinate one heterozygous red-flowering pea plant. This is equivalent to cross-pollinating two heterozygous red-flowering pea plants.

In symbols:

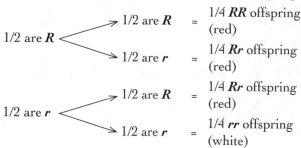

Genotypic ratio expected in the offspring generation is 1 RR : 2 Rr : 1 rr (1/4 RR, 1/2 Rr, 1/4 rr).

Phenotypic ratio expected in the offspring generation is 3 red : 1 white (3/4 red, 1/4 white).

This process of working the problem could be abbreviated by writing the information in a Punnett square:

		possible genotypes of gametes from other parent	
		1/2 R	1/2 r
possible genotypes of gametes from one parent	1/2 R	1/4 RR offspring	1/4 Rr offspring
	1/2 r	1/4 Rr offspring	1/4 rr offspring

14

Dominance vs. Partial Dominance

The allele for red flowers is the normal functional version of the gene that encodes a protein called an enzyme that is required for the production of red pigment. The allele for white flowers is a mutant form of this gene, encoding a nonfunctional version of the required enzyme.

Plants heterozygous for flower color (*Rr*) have only one copy of the allele for red flowers. In heterozygous *garden peas*, that one copy of the *R* allele can be used efficiently enough to make enough protein such that the cell can synthesize enough red pigment to make the flower petals appear red. In heterozygous *snapdragons*, however, that one copy of the *R* allele does *not* result in enough protein to produce the normal level of synthesis of red pigment; thus, heterozygous snapdragon petals produce less than the normal amount of red pigment and appear *pink*. In snapdragons, the *R* allele shows **partial dominance**, making the heterozygote a *distinctive phenotype* that is different from the phenotypes of the *RR* and *rr* homozygotes. In the case of partial dominance, allele symbols such as *R* and *R'* are generally used because neither allele is fully recessive and therefore has not earned the lower case letter as its symbol.

Q1. What is the expected phenotypic ratio in the offspring generation when you cross two pure-breeding (homozygous), red-flowering *pea* plants? Show your work.

Q2. Would you ever get any white-flowering *pea* plants if you cross a homozygous red-flowering pea plant and a heterozygous red-flowering pea plant? Show your work.

Q3. What is the expected phenotypic ratio in the offspring generation when you cross a red and a white *snapdragon*? Show your work.

Q4. What is the expected phenotypic ratio in the offspring generation when you cross a pink and a white *snapdragon*? Show your work.

Q5. What is the expected phenotypic ratio in the offspring generation when you cross two pink *snapdragons*? Show your work.

150

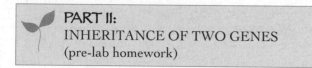
Two different genes may be monitored during a genetic cross. For example, the gene for flower color and the gene for plant height may be of interest to the researcher. Gregor Mendel found, in modern terminology, that *a gamete contains one copy of each kind of gene* and that *two different genes assort independently* of each other into the gamete (Mendel's Second Law of Heredity, or the Law of Independent Assortment). This means that assortment of the two copies of one gene into the gametes is not influenced by the manner of assortment of the two copies of the other gene.

Peas:

R =	red-flowering, dominant	*T* =	tall, dominant
r =	white-flowering, recessive	*t* =	short, recessive

Each gamete has | one copy of the gene for flower color | *AND* | one copy of the gene for plant height |

The *rrtt* parent produces gametes:
All have *r* and all have *t*
= all gametes are genotype *rt*.

The *RrTT* parent produces gametes:
1/2 have *R* and all have *T*
= 1/2 of gametes are genotype *RT*.
1/2 have *r* and all have *T*
= 1/2 of gametes are genotype *rT*.

The *RrTt* parent produces gametes:
1/2 have *R* and 1/2 of these have *T*
= 1/4 of gametes are *RT*.
. . . and 1/2 of these have *t*
= 1/4 of gametes are *Rt*.
1/2 have *r* and 1/2 of these have *T*
= 1/4 of gametes are *rT*.
. . . and 1/2 of these have *t*
= 1/4 of gametes are *rt*.

Example 3

Cross a purebreeding (homozygous) red-flowering tall with a white-flowering short pea.

Written as genotypes, this is:

$$\begin{array}{ccc} RRTT & \times & rrtt \\ \text{(red, tall)} & & \text{(white, short)} \end{array}$$

gametes: all are **RT**   all are **rt**

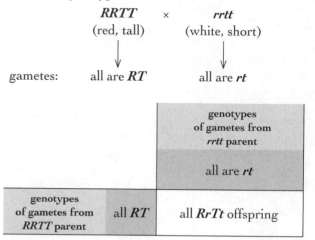

Expected genotypic ratio in offspring generation:
all are **RrTt**.

Expected phenotypic ratio in offspring generation:
all are red-flowering, tall.

Example 4

Perform the cross:

$$\begin{array}{ccc} RrTT & \times & rrtt \\ \text{(red, tall)} & & \text{(white, short)} \end{array}$$

gametes: 1/2 are **RT**   all are **rt**
            1/2 are **rT**

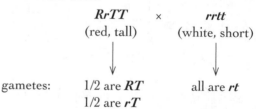

Expected genotypic ratio in offspring generation:
1 **RrTt** : 1 **rrTt**

Expected phenotypic ratio in offspring generation:
1 red-flowering, tall : 1 white-flowering, tall.

14

Example 5

Perform the cross:

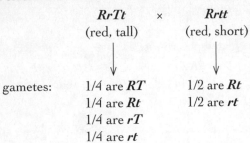

$$RrTt \quad \times \quad Rrtt$$
(red, tall)   (red, short)

gametes:
1/4 are **RT**   1/2 are **Rt**
1/4 are **Rt**   1/2 are **rt**
1/4 are **rT**
1/4 are **rt**

Think about the fact that the *RrTt* parent produces an *RT* gamete 1/4 of the time, and this gamete is fertilized half the time by an *Rt* gamete of the other parent, producing the genotype *RRTt* in the offspring 1/8 of the time (one-half of a quarter is 1/8, or mathematically expressed $1/4 \times 1/2 = 1/8$).

		genotypes of gametes from *Rrtt* parent	
		1/2 **Rt**	1/2 **rt**
	1/4 **RT**	1/8 **RRTt**	1/8 **RrTt**
	1/4 **Rt**	1/8 **RRtt**	1/8 **Rrtt**
genotypes of gametes from *RrTt* parent	1/4 **rT**	1/8 **RrTt**	1/8 **rrTt**
	1/4 **rt**	1/8 **Rrtt**	1/8 **rrtt**

Expected genotypic and phenotypic ratios in offspring generation:

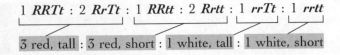

1 *RRTt* : 2 *RrTt* : 1 *RRtt* : 2 *Rrtt* : 1 *rrTt* : 1 *rrtt*

3 red, tall : 3 red, short : 1 white, tall : 1 white, short

Q6. Your turn now! Do the cross:

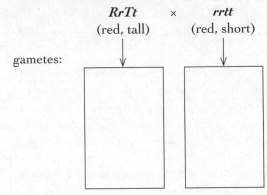

$$RrTt \quad \times \quad rrtt$$
(red, tall)   (red, short)

gametes:

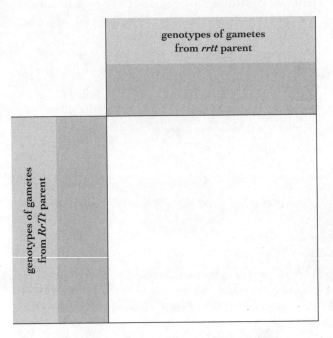

	genotypes of gametes from *rrtt* parent
genotypes of gametes from *RrTt* parent	

Phenotypic ratio expected in the offspring generation from this cross:

Q7. Cross two red-flowering, tall pea plants, each heterozygous for both traits.

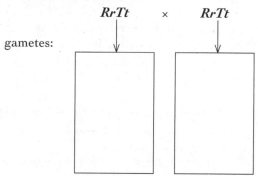

gametes:

genotypes of gametes from one parent	genotypes of gametes from other parent			

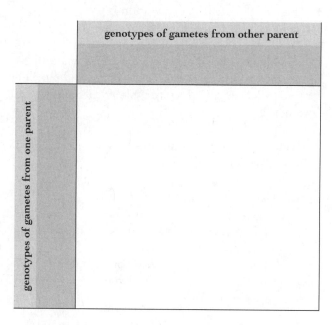

Phenotypic ratio expected in the offspring generation from this cross of pea plants:

Phenotypic ratio expected in the offspring generation if this *RrTt × RrTt* cross had been done in snapdragons where red-flowering is partially dominant to white:

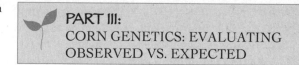

PART III:
CORN GENETICS: EVALUATING OBSERVED VS. EXPECTED

NOTE: The investigations in Part III includes an opportunity to practice the chi-square test—a statistical test used in various fields of study to evaluate data.

BACKGROUND

The tassel at the top of the corn plant contains the male flowers. The ear of corn consists of a modified lateral branch that is covered with many female flowers and encased in modified leaves called the corn husks. Tightly clasping corn husks as well as kernels that do not drop at maturity are features that have been bred into domesticated corn thousands of years ago.

The pistil of a female flower has an ovary containing one ovule, within which develops one egg cell. The pistil's ovary bears one thin style, called corn silk, that elongates and emerges from the end of the husks. The emerged portion of the silk is the stigma, which continues to elongate and is receptive to pollen anywhere along its emerged length. Corn is wind pollinated. For details on corn reproduction, see the website "SEX in the Corn Field: How's It Done?" at http://www.agry.purdue.edu/ext/corn/pubs/corn-02.htm. (If the address has changed, search for the site's name.)

Double fertilization occurs in all flowering plants, including corn. Each ovule develops a cluster of eight haploid nuclei, which include the egg cell and two "polar nuclei" that are genetically identical to the egg. Each pollen grain contains two sperm cells, each of which engages in a separate fertilization:

- 1 sperm + 1 egg = diploid zygote (2n) that grows into the diploid plant **embryo** and its attached cotyledon in the seed
- 1 sperm + 2 polar nuclei = triploid cell (3n) that grows into **endosperm** (embryo's nutritive tissue)

14

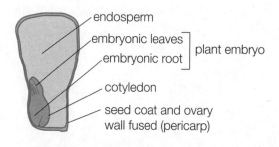

Figure 14.2 Corn kernel

The **kernel** (see Figure 14.2) is botanically a dry fruit—it is the mature ovary wall containing one seed. Note that in grasses such as corn, the maturing ovary wall and the developing seed coat grow together into one enveloping sheet, the **pericarp** (the stuff that gets stuck between your teeth when eating corn). The **aleurone** is the outermost protein-rich layer of cells of the endosperm, just beneath the pericarp.

Genetics of corn kernel parts:
* The pericarp (ovary wall and seed coat) is maternal tissue; it is diploid (two copies of each gene).
* The embryo and its attached cotyledon are the result of the fertilization of egg and sperm; it is the diploid offspring plant.
* The starchy endosperm and its outermost protein-rich aleurone layer are the result of the second fertilization; these parts are triploid, having three copies of each kind of gene—one from the father's sperm and two from the mother's two polar nuclei.

The overall color of a corn kernel is determined by the sum effect of genes affecting:
* The **pericarp**—colorless or colored
* The **aleurone**—colorless or colored; a colored aleuron is visible through a colorless pericarp.
* The **endosperm**—colorless or colored; the endosperm color is visible through colorless pericarp and colorless aleurone.

Why has corn been a favorite genetics research organism? Each corn kernel contains a separate offspring. An ear of corn is a compact arrangement of several hundred offspring—conveniently providing a large sample size for the researcher to evaluate when studying the inheritance of genes affecting kernel color or texture.

Note that the maternal pericarp tissue must be *colorless* for the underlying aleurone or endosperm color to be easily visible. Carolina Biological Supply Company and Ward's provide corn ears that have colorless pericarps; the colors that you see through each colorless pericarp are of the aleurone layer and/or the endosperm—both of which result from the second fertilization.

Here are the symbols for, and effects of, five of the genes affecting corn kernels:

P = purple aleurone (encodes an enzyme required for converting red into purple pigment)
p = red aleurone (mutated form of the gene encodes nonfunctional enzyme)

C = enables red or purple pigment to be made from the previously listed gene (activates some steps in the synthesis of red and purple)
c = no color in aleurone (recessive mutation; unable to activate these steps)
C^I = no color in aleurone (mutation is dominant to allele C; prevents activation of these steps)

R = enables red or purple pigment to be made (activates some steps in the synthesis of red and purple)
r = no color in aleurone (unable to activate these steps)

Y = yellow endosperm (encodes an enzyme required for synthesis of yellow carotenoid pigment)
y = white endosperm (mutated nonfunctional enzyme)

S = plump kernels when dry (starchy endosperm)
s = wrinkled translucent kernels when dry (altered starch structure in endosperm contains more water when fresh; loses more water volume in drying and thus becomes wrinkled)

III.A. ONE GENE—OBSERVED VS. EXPECTED PHENOTYPES

MATERIALS

* segregating corn ears, one gene involved (for each student pair, a corn ear in a plastic bag labeled with the parental cross)
  - *Rr* × *Rr* (Carolina HT-17-6500)
  - *Rr* × *rr* (Carolina HT-17-6502; Ward's 86 W 8900)
  - *Cc* × *cc* (Carolina HT-17-6522)
  - *Su* × *Su* (Carolina HT-17-6540)
  - *Cc* × *Cc* (Ward's 86 W 8910)

INVESTIGATION

☐1. Record the genetic cross that produced your corn ear: _____

On your corn ear, count the *number* of offspring (kernels) in each phenotypic class.

a. Your counts:

b. Total number of kernels counted: _____

☐2. Work through the genetic cross that produced your corn ear, and determine the phenotypic ratio expected in the offspring generation on your corn ear.

☐3. Use the expected ratio to calculate the *number* of offspring (kernels) expected in each phenotypic class for your total number of kernels counted (see Appendix 3, Part III, "Chi-Square Test for Genetics Problems").

☐4. Are your observed data close enough to the expected numbers to support the hypothesis that this corn ear was the result of the indicated cross? To determine this, follow the directions in Appendix 3, Part III, to perform the chi-square statistical test and draw a conclusion. Show your work.

☐5. You may be instructed to trade corn ears with another student pair, repeat the analysis, and check each other's results.

III.B. TWO GENES—OBSERVED VS. EXPECTED PHENOTYPES

MATERIALS

* segregating corn ears, two genes involved (for each student pair, a corn ear in a plastic bag labeled with the parental cross)
  - *RrSu* × *RrSu* (Carolina HT-17-6600)
  - *SuYy* × *SuYy* (Ward's 86 W 8917)
  - *RrSu* × *rrsu* (Carolina HT-17-6602)
  - *C'CSu* × *C'CSu* (Ward's 86 W 8915)

INVESTIGATION

☐1. Record the genetic cross that produced your corn ear: _____

On your corn ear, count the *number* of offspring (kernels) in each phenotypic class.

a. Your counts:

b. Total number of kernels counted: _____

☐2. Work through the genetic cross that produced your corn ear, and determine the phenotypic ratio expected in the offspring generation on your corn ear.

parental genotypes: _____ × _____

gametes:

genotypes of gametes from other parent

genotypes of gametes from one parent

Phenotypic ratio expected:

☐3. Use the expected ratio to calculate the number of offspring (kernels) expected in each phenotypic class for your total number of kernels counted (see Appendix 3, Part III, "Chi-Square Test for Genetics Problems").

☐4. Are your observed data close enough to the expected numbers to support the hypothesis that this corn ear was the result of the indicated cross? To determine this, follow the directions in Appendix 3, Part III, to perform the chi-square statistical test and draw a conclusion. Show your work.

☐5. You may be instructed to trade corn ears with another student pair, repeat the analysis, and check each other's work.

III.C. ADVANCED: TWO GENES INTERACTING

MATERIALS

- segregating corn ears, two genes involved (each ear in a plastic bag—labeled with its parental cross)
  - $C^ICYy \times C^ICYy$
    note that both parents are also PP and RR
    (Carolina HT-17-6680)
  - $CcRr \times CcRr$
    note that both parents are also pp and yy
    (Carolina HT-17-6690)
  - $C^ICRr \times C^ICRr$
    note that both parents are also PP and YY
    (Carolina HT-17-6700; Ward's 86 W 8920)

INVESTIGATION

☐1. Record the genetic cross that produced your corn ear: _____
On your corn ear, count the *number* of offspring (kernels) in each phenotypic color class.
  a. Your counts:

  b. Total number of kernels counted: _____

☐ 2. Work through the genetic cross that produced your corn ear:

parental genotypes: _____ × _____

gametes:

genotypes of gametes from other parent

genotypes of gametes from one parent

☐ 3. Write the expected genotypic ratio below. Now check the background section of Part III for the meaning of the gene symbols, and determine the phenotype of each class in the genotypic ratio; finally, combine those classes having the same phenotype to produce the overall phenotypic ratio expected in the offspring generation. (*Be careful!* Note that corn must have both C and R alleles present in order for P or p alleles to produce any color. Also remember that C^I is dominant to C.)

☐ 4. Use the expected phenotypic ratio to calculate the expected *number* of offspring (kernels) in each phenotypic class, as explained in Appendix 3, Part III, "Chi-Square Test for Genetics Problems."

☐ 5. Are your observed data close enough to the expected numbers to support the hypothesis that this corn ear was the result of the indicated cross? To determine this, follow the directions in Appendix 3, Part III, to perform the chi-square statistical test and draw a conclusion. Show your work.

☐ 6. You may be instructed to trade corn ears with another student pair, repeat the analysis, and check each other's results.

14

III.D. ADVANCED: INDIAN CORN FROM UNKNOWN GENETIC CROSS

MATERIALS

• ears of Indian corn with 2–4 phenotypic classes per ear (from grocery store or farmers market)

NOTE: Because corn has many genes that affect the kernel's color as well as the color's intensity and pattern, certain crosses can produce ears having numerous different phenotypic classes.

INVESTIGATION

□1. On a corn ear, count the number of offspring (kernels) observed in each phenotypic class.
    a. Your counts:

    b. Total number of kernels counted: _____

□2. In earlier parts of this lab, you have seen various phenotypic ratios produced by various crosses. Which of these standard *phenotypic* ratios do your counts approximate?

From this, *hypothesize the parental genotypes* that were crossed to produce your corn ear; use the symbols defined at the beginning of Part III.

_____ × _____

□3. Work through this hypothesized genetic cross, and calculate the *numbers* of offspring (kernels) expected in each phenotypic class for your total number of kernels counted.

□4. Do the chi-square test to determine whether your observed counts in the phenotypic classes are close enough to the calculated expected numbers so that you can accept the hypothesized cross as having produced this ear of corn.

	Most individuals	Mutant plant 1	Mutant plant 2
Genotype	*DDHH*	*D∂HH*	*DDHb*
Phenotype	drought & heat sensitive	drought & heat sensitive	drought & heat sensitive
Asexual reproduction: For example, a shoot breaks off and roots, forming a new plant that is genetically identical to the original plant.			
Genotypes after asexual reproduction			
Phenotype after asexual reproduction			
Selfing sexual reproduction: Each plant self-pollinates.			
Genotypes in the population after a generation of selfing			
Phenotypes in the population after a generation of selfing			
Outcrossing sexual reproduction			

Cross mutant 1 with mutant 2: What is the genotypic ratio and phenotypic ratio in the first offspring (F_1) generation?

Can certain F_1 genotypes be crossed to produce an offspring generation that includes individuals that are **both** drought and heat tolerant? Show your work.

14

Table 14.1 Asexual reproduction vs. selfing vs. outcrossing

PART IV:
FURTHER INVESTIGATIONS

IV.A. ASEXUAL REPRODUCTION VS. SELFING VS. OUTCROSSING: EFFECT ON PHENOTYPIC DIVERSITY?

The greater the diversity of phenotypes present in a species, the greater chance there is for long-term survival of the species. When the environment changes over time, a species with a good diversity of phenotypes will be likely to have some individuals that are adapted to the new conditions; these individuals will be able to survive and perpetuate the species.

Suppose that a particular plant species occurs in a moist, cool area. These plants are genotype *DDHH*, where dominant allele *D* of the gene for moisture requirement makes the plant sensitive (intolerant) to drought and dominant allele *H* of the gene for temperature requirement makes the plant sensitive (intolerant) to heat. Suppose that due to a mutation, one plant is *D∂HH* with the recessive allele *∂* for drought tolerance. Suppose another mutant plant is *DDHh* with the recessive allele *h* for heat tolerance.

All these plants (*DDHH, D∂HH, DDHh*) have the phenotype of requiring a moist, cool environment. As global warming continues over the generations, individuals in the plant population that have the phenotype of both drought and heat tolerance would have a survival advantage.

Q8. Examine the effect of *asexual reproduction* versus *selfing* versus *outcrossing* on the phenotypic diversity in this population after a generation or two. Fill in Table 14.1.

Q9. Does self-fertilization produce offspring that are all phenotypically identical with the parent?

Would self-fertilization of this population over several generations ever produce individuals that are both drought and heat tolerant?

Q10. Would the species have the best chance of surviving global warming if it reproduces by asexual reproduction or by selfing or by outcrossing? Explain .

IV.B. ADVANCED: DOSAGE EFFECT IN ENDOSPERM

Have you noticed that on some ears, the kernels in the "yellow phenotype" class were various shades—deep yellow, light yellow, or pale yellow? In this part, you will explore the reason for this.

MATERIALS

- segregating corn ears: *Yy* × *Yy* (Ward's 86 W 8912), one ear per plastic bag

INVESTIGATION

☐1. For the cross *Yy* × *Yy*, determine the genotypic ratio of the **offspring embryonic plants** in the kernels.

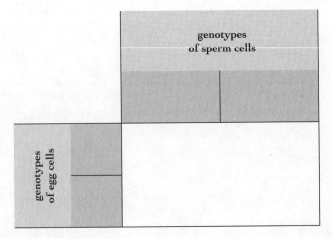

Genotypic ratio expected in the offspring embryonic plants:

160

☐ 2. For the cross $Yy \times Yy$, determine the genotypic ratio for the kernel's **triploid endosperm** that is produced by fusion of a paternal sperm cell and two maternal polar nuclei. Remember that each polar nucleus is genetically identical to the egg cell in its ovule. So, if the egg cell contained allele Y, then each polar nucleus contains an allele Y and thus the pair of polar nuclei contribute two copies of allele Y.

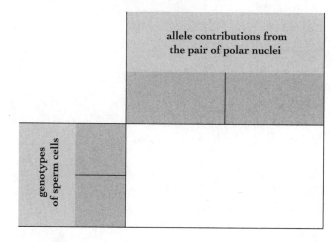

Genotypic ratio expected in the offspring endosperms:

☐ 3. The yellow kernels look yellow because of the yellow pigment produced in the endosperm. Why would some kernels look deep yellow, others light yellow, and yet others pale yellow? Explain.

LITERATURE CITED

For more details and pictures of corn reproduction:

Nielsen, R. L. n.d. "SEX in the Corn Field: How's It Done?" http://www.agry.purdue.edu/ext/corn/pubs/corn-02.htm (accessed February 19, 2006). West Lafayette, IN: Department of Agronomy, Purdue University.

For detailed information on corn genes and mutations:

Neuffer, M. G., E. H. Coe, and S. R. Wessler. 1997. *Mutants of Maize*. Cold Spring Harbor, NY: Cold Spring Harbor Laboratory Press.

14

LAB 15

DIVERSITY: ALGAE, FUNGI, LICHENS

I. Diversity in the field
II. Up close and personal: algae, fungi, and lichens under the microscope
III. Supermarket survey: carrageenan, agar, alginates

NOTE: The objectives are given at the beginning of each part of this lab.

BACKGROUND

A lichen is just a fungus that has discovered agriculture.

— lichenologist Trevor Goward, University of British Columbia

Under the microscope, a typical **lichen** in cross section looks like spaghetti and meatballs—within the tangle of fungal strands are the round photosynthetic green algal and/or cyanobacterial cells. A lichen is not one organism, but rather a **symbiotic** association—different organisms living together (Greek: *symbiosis* = a living together). Note, however, that a lichen is classified and given a species name as if it were a single organism.

Fungi (plural of *fungus*) are important decomposers, although some are pathogens. Fungal strands secrete certain enzymes to break down the **substrate** (the dead or living material on which the fungus grows); the organic breakdown products are very efficiently absorbed and then further broken down by cellular respiration to supply the fungus with energy. Fungi are also highly efficient at absorbing minerals and water from the substrate as well as from the atmosphere.

Algae and **cyanobacteria** (plural of *alga* and *cyanobacterium*) receive their energy by photosynthesis. Some kinds of cyanobacteria can, in addition, fix nitrogen—convert the stable nitrogen gas into usable nitrogen products.

In a lichen, the symbiotic association between the **mycobiont** (the fungus) and the **photobiont** (the photosynthetic alga and/or cyanobacterium) is mutually beneficial. The photobiont benefits by receiving minerals, water, and some carbon dioxide from the fungus; in addition, the photobiont is protected from intense sunlight by the often pigmented fungal strands that become opaque upon desiccation. The fungal **mycobiont** benefits by receiving substantial amounts of organic photosynthetic products along with oxygen from the photobiont; when the photobiont is a nitrogen-fixing cyanobacterium, the fungus also receives usable nitrogen products. This mutually beneficial association allows lichens to survive and thrive in exposed, intensely sunny, nutrient-poor environments.

Other fungi engage in a different kind of symbiotic relationship—these **mycorrhizal** fungi form a mutually beneficial association with the roots of specific kinds of plants (Greek: *mykes* = fungus, *rhiza* = root). The mycorrhizal fungus penetrates and absorbs minerals and water from a much larger volume of soil than the root system could reach. The plant receives some of the minerals and water absorbed by the fungus, while the fungus receives organic photosynthetic products from the plant roots.

15

NOTE: Your instructor will indicate the parts of this lab that your class will perform.

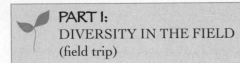

PART I:
DIVERSITY IN THE FIELD
(field trip)

I.A. FUNGI AND LICHENS: DIVERSITY AND SUBSTRATE SPECIFICITY (field trip)

OBJECTIVES

1. Observe the diversity of fungi and lichens in your area.
2. Determine the preferred substrate for the various kinds of fungi and lichens observed.
3. Consider the factors that may cause certain species of fungi or lichens to grow on certain substrates and/or be associated with certain trees.

BACKGROUND

The two largest groups of fungi are the **sac fungi** (Ascomycetes) and the **club fungi** (Basidiomycetes). Most members of these two groups grow vegetatively (nonreproductively) as fungal strands until appropriate conditions of moisture, temperature, and so forth induce them to form sexual spores, often in special fruiting bodies. Those fruiting bodies that consist of a stalk and cap are commonly called mushrooms. Mushrooms and other fruiting bodies of fungi can be found on the ground and on living or dead trees. The fruiting bodies of a mycorrhizal fungus will be found near its symbiotic plant.

Among the lichens, there are three main body types (see Figure 15.1). The **crustose** lichens form thin crusts growing tightly against the substrate; the **foliose** lichens are flat sheets with distinct margins, wrinkled to various degrees; the **fruticose** lichens grow away from the substrate, as upright or hanging stalks or bushy structures.

Lichens commonly occur on rocks, on the ground, and on living and dead trees. Lichens are not common in urban areas, because the fungal component is efficient at absorbing not just nutrients, minerals, and water but also toxic pollutants. The lichen diversity in an area is a good indicator of air quality there.

For each of the fungi and lichens that you find during the field trip, you will determine whether it shows a preference for a particular substrate and whether it is associated with particular trees.

MATERIALS

- nearby natural area, on or off campus
- field guide to trees (such as Brockman 2001; see Lab 17, "Plant Diversity: Seed Plants," for an annotated bibliography of tree guides)
- field guide to mushrooms (such as McKnight and McKnight 1987)

INVESTIGATIONS

NOTE: When outdoors, be alert for venomous animals as well as poison ivy, nettles, and other such plants.

☐1. At a nearby natural area, your instructor will point out several fungi and lichens, to get you started in recognizing these highly diverse structures. (Your instructor may identify some of the common species; unidentified species may be kept track of by giving them numbers or making up names for them.) Students may rotate the job of being the identifier of nearby trees.

Figure 15.1 Lichen body types: (a) crustose lichen; (b) foliose lichen; (c) fruticose lichen

The format of your field trip may be one of the following, depending on the condition of your natural area:

- The instructor may identify a certain set of species; then the students can spread out to look for and note the locations of other individuals of those species.
- The instructor may lead the group down a trail, requiring students to call a halt whenever they have found a fungus or lichen to examine.
- After the first observation of a particular species, the instructor may assign that species to a particular student, who is then to focus on finding other individuals of that species.

☐2. Record each species observed, and its substrate (Worksheet 1); then carefully observe and note characteristics of its location such as:

- On the ground: level or sloping; on bare soil, leaf litter, conifer needles, hardwood mulch; near or under which species of trees
- On rock: top or sides of rock; vertical rock face, facing which direction
- On tree: trunk, branch, or crotch; which species of tree; how high on the tree
- On rotting log: which tree species; on the sides or ends of the log
- Light condition: sun or shade
- Moisture condition: well drained or moist

☐3. Each time you find another site for a particular species, either record the characteristics of the new site if it is different from the first, or put a check mark after the existing description if it is the same.

☐4. At the end of the trip: Draw conclusions about the preferred substrate of each kind of fungus and lichen.

☐5. Look up the identified species in reference books to compare your observations to the published statements concerning the substrate preferences.

WORKSHEET 1 : Fungi and Lichens—Diversity and Substrate Specificity

Species of fungus or lichen	Location: on the ground, rock, tree, or log	Characteristics of that location

15

Q1. For which species do your observations agree with the published statements?

Q2. If your observations do not agree with the published statements, does that make your observations wrong? Discuss.

Q3. Were any fungi consistently found under or near certain tree species? Explain why such an association might exist between a fungus and a tree.

Q4. If certain species of lichen were found on the bark of only certain kinds of trees, why might that be?

Q5. Were lichens typically distributed evenly over a tree's bark or primarily in certain locations on the tree? Why might this be?

I.B. ALGAL DIVERSITY: COMPARING TWO SITES (plankton tows)

OBJECTIVES

1. Learn how to take a plankton tow.
2. Through critical observation (without getting into scientific identification of species), distinguish the different kinds of algae in a sample and determine whether one or more kinds are predominant.
3. Compare plankton tows from two different locations and determine whether the species composition and relative abundance is noticeably different in the two locations.
4. Realize the importance of being able to identify algae when studying the effect of pollutants on aquatic ecosystems.

BACKGROUND

To get a concentrated sample of the algae in a body of water, scientists use a plankton net. To take a plankton tow from moving water, the plankton net is placed into the water current for a while. To sample the algae from a still body of water, the plankton net is pulled through the water—such as by pulling it along a dock or behind a boat, or by tossing it out and pulling it toward shore. Note that you will likely also collect various small invertebrates, silt, and bits of organic debris in the tow.

In this lab, you will compare and contrast the algal community of two different waters.

MATERIALS

- plankton net, 50 μm mesh, with 2 collecting bottles (Carolina HT-65-2158); one net per student group
- long, stout nylon or other cord, to attach to the plankton net
- pipets with bulbs
- compound microscope per student, lens paper, microscope slides, and cover slips
- latex gloves: small, medium, and large

INVESTIGATIONS

CAUTION: Cyanobacteria, formerly called bluegreen algae, often are bluegreen in color and may proliferate ("bloom") to the point of turning the body of water greenish. Some cyanobacteria produce very potent toxins that damage the nervous system, liver, or skin.

Dinoflagellates can also proliferate to the point of causing the water to become variously colored; some dinoflagellates also produce very potent toxins. Therefore, if the water has a color or if that water has a history of periodically having color, it is safest not to touch it with bare hands; wear latex gloves.

☐1. Work in groups of four or more. Decide on the question to investigate, such as one of the following:

- Is the algal community of two different locations in the same body of water the same?
- Is the algal community in two different streams the same?
- Is the algal community in a stream the same as in a pond?
- Is the algal community upstream and downstream from a pollution source the same?
- Is the algal community upstream and downstream from a sewage treatment plant the same?
- Is the algal community in two different tanks (for fish, turtles, aquatic plants, etc.) the same?
- What effect does net mesh size (50 µm versus 150 µm mesh) have on the algal diversity in tows taken from the same site?

☐2. Record your question and hypothesis on Worksheet 2.

☐3. Decide on your procedure—where, how, and for how long you will take a sample. Record your procedure on Worksheet 2.

☐4. Attach a bottle firmly to the end of the plankton net. Take a plankton tow of a site by placing the net into the current for a set amount of time or by pulling it through the water a set number of times. When sampling a tank, scoop out a set amount of water and pour it into the net.

☐5. Hold the bottle and net upright to drain the excess water; you may speed up this process by shaking the bottle sideways to slosh the water around in the net a bit. When the net is drained, remove the bottle and cap it.

☐6. Rinse the inside of the net and attach another bottle; take a plankton tow of the second site, using the same procedure as for the first site.

☐7. Refer to Lab 2 "Microscopy and Plant Cells" for instructions and prepare several microscope slides of each sample; view under a compound microscope, starting at a corner of the cover slip and moving systematically back and forth across the entire width of the cover slip to view all areas.

☐8. For each slide, record the different kinds of algae present and their abundance. Consult Figure 15.2 for a few common types of algae; you will not, however, have the training to correctly identify all the algae that you see. You can keep track of the unidentified kinds by drawing them and creating names for them. You can describe the abundance of a particular type of alga as the number of individuals per slide, or per single pass across the cover slip, or per field of view under a particular objective.

☐9. Pool the observations of all members of the group. Based on your observations, determine the similarities and differences between the two sites; record your results on Worksheet 2.

Q6. In your two sampled sites, what similarities are there in algal composition?

Q7. In your two sampled sites, how is the algal composition different? What factors might be causing the difference in species composition?

Q8. The rather common but *unusually tiny* marine algae called coccolithophorids were not discovered until relatively recently (a little over a century ago). Why do you suppose they were so belatedly discovered?

15

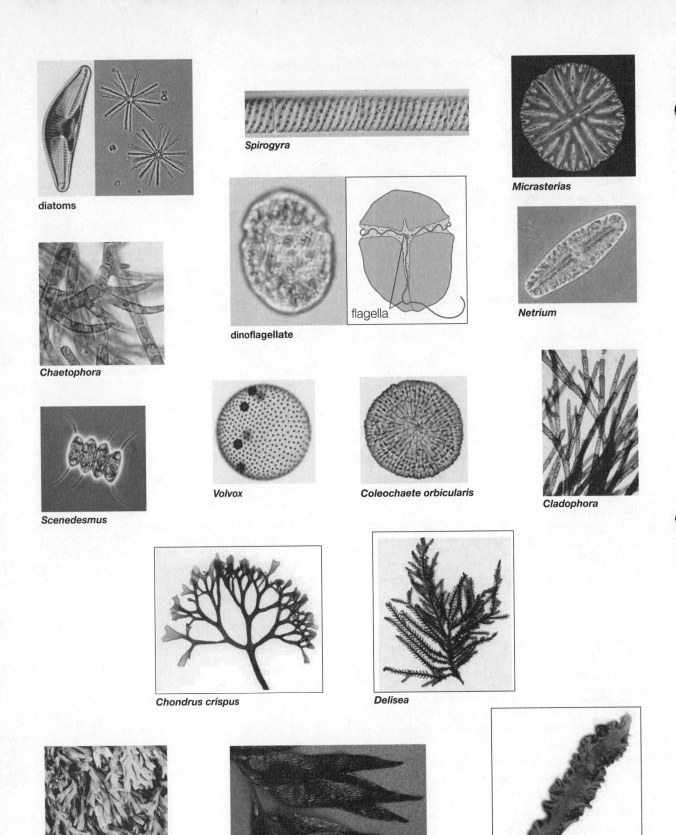

diatoms

Spirogyra

Micrasterias

Netrium

Chaetophora

dinoflagellate

flagella

Scenedesmus

Volvox

Coleochaete orbicularis

Cladophora

Chondrus crispus

Delisea

Fucus

Macrocystis

Laminaria

Figure 15.2 Algal diversity

WORKSHEET 2: Algal Diversity

Your question:

Your hypothesis:

Sites selected: 1._____ 2._____

Procedure:

Algal type	Abundance in site 1		Abundance in site 2	
	Your data	Pooled data	Your data	Pooled data

15

PART II:
UP CLOSE AND PERSONAL: ALGAE, FUNGI, AND LICHENS UNDER THE MICROSCOPE

OBJECTIVES

1. Understand the basis of classifying a fungus as an ascomycete or basidiomycete.
2. Recognize asci and basidia under the microscope.
3. Know where asci and basidia are located on various ascocarps and basidiocarps.
4. Recognize the ascocarps on a lichen, and understand their origin and function.
5. Be able to distinguish cross sections of an ascocarp, basidiocarp, and lichen under the microscope.
6. Become familiar with the great diversity of form in the algae.
7. Be able to recognize diatoms, dinoflagellates, desmids, *Fucus*, *Laminaria*, *Spirogyra*, *Volvox*, and/or others specified by your instructor.

BACKGROUND

The Kingdom Fungi is divided into groups called Divisions or Phyla based on reproductive features. The two largest Phyla are the ascomycetes and the basidiomycetes. In the ascomycete fungi (sac fungi), meiosis cell division produces sexual spores that are enclosed in a sac called an **ascus** (Greek: *askos* = bladder) (see Figure 15.3a). The asci (plural of ascus) containing their **ascospores** are tightly packed together in or on the **ascocarp** fruiting structure; asci are on the upper surface of the scarlet cup fungus, inside a truffle, and lining the pits on the cap of a morel (see Figure 15.4). Mature asci typically explode, tossing the ascospores into the air. In the basidiomycete fungi (club fungi), meiosis occurs in a club-like cell called a **basidium**, and the resulting sexual spores are budded off from this basidium (Greek: *basis* = base) (see Figure 15.3b). These basidia (plural of *basidium*) and their **basidiospores** are produced in or on the **basidiocarp** fruiting structure— inside the puffballs, inside the "bird eggs" of the bird's nest fungus, in the gelatinous smelly material at the end of a stinkhorn, on the surfaces of the curtain-like gills hanging down beneath the cap of the Portabella mushroom, and on the walls of the tubes extending vertically up into the shelf fungus from each pore on its lower surface (see Figure 15.4). The corn smut and wheat rust fungi are basidiomycetes because they do produce basidiospores, although they do not produce the basidiocarp fruiting structures. Baker's yeast is an ascomycete because it produces asci, although not ascocarps.

Note that an ascomycete or basidiomycete can grow for years as vegetative microscopic fungal strands, degrading the material on which they grow, before conditions are appropriate to induce the formation of a fruiting structure. A mushroom or shelf fungus is only the fruiting structure, not the entire organism.

Crustose and foliose lichens (see Figure 15.1), which we discussed at the beginning of this lab, frequently have structures on their upper surface that look like tiny squashed jelly doughnuts. These are ascocarps—

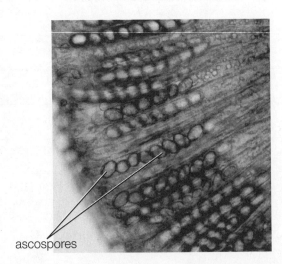

(a) many linear asci containing ascospores

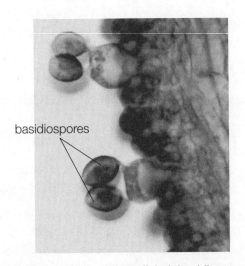

(b) two basidia budding off their basidiospores

Figure 15.3 Ascomycete versus basidiomycete

the fruiting bodies produced by the lichen's ascomycete fungus. Some fruticose lichens such as British soldier lichen have colorful ascocarps at their branch tips. A cross section through these ascocarps shows the densely packed asci.

The term *algae* is an informal name for the various protist groups that are usually photosynthetic. Algae are generally aquatic organisms that come in a great diversity of shapes and sizes, ranging from unicellular to filamentous to colonial to multicellular bodies (see Figure 15.2). A *diatom* consists of a cell encased in "a little glass box" made of silica, having a base and a lid, much like a petri dish. *Coralline* algae deposit calcium carbonate in their cell walls, making a stony mass that adds to the structure of coral reefs. The multicellular, macroscopic *kelp* seaweeds, such as *Laminaria*, are tough and rubbery, and can be more than 60 meters long. Roughly half of the *dinoflagellates* are photosynthetic; the dinoflagellates are unicellular and have two whiplike flagellae—one encircles the cell and the other is longitudinal; the beating of the two flagellae causes the cell to spin while moving forward. Although many algae are various shades of green, other algae are shades of red, brown, or gold.

MATERIALS

- Specimens may be set up at stations around the lab room. Listed microscope slides are mounted on compound microscopes at the station.

Station 1:
- *Morchella*, asci, sect. (Ward's 91 W 2462)

Station 2:
- *Coprinus* mushroom, c.s. (Carolina HT-29-8176, Ward's 91 W 3211)

Station 3:
- three dissecting microscopes with crustose, foliose, and fruticose lichens (locally available or Carolina HT-15-6400, Ward's 85 V 5915)

Station 4:
- Lichen thallus, sect. (Carolina HT-29-8476, Ward's 91 W 3951)
- one-sided razor blade, a portion of hydrated foliose lichen, forceps, and a dissecting microscope

Station 5:
- two "unknown" slides, such as the following, with their labels covered with numbers:
  Peziza, apothecium, sect. (Carolina HT-29-7980, Ward's 91 W 2480)
- Lichen ascocarps, sect. (Carolina HT-29-8488, Ward's 91 W 3952)

Station 6:
- locally collected bird's nest fungus, puffball, or other fungi as available

Station 7:
- dissecting microscopes and shelf fungi showing various pore sizes, from easily visible to not visible to the naked eye

Station 8:
- several compound microscopes (one per culture)
- lens paper, microscope slides, cover slips, and pipets with bulbs
- several living cultures of algae (labels covered with numbers) such as: *Cladophora, Micrasterias, Navicula, Peridinium, Scenedesmus, Spirogyra, Volvox* (Carolina HT-15-2105, -2345, -3045, -3290, -2510, -2525, -2665; Ward's 86 W 0145, 0270, 1210, 2900, 0600, 0650, 0805 respectively)

Station 9:
- *Fucus, Laminaria, Chondrus*, labeled with numbers (if not locally available, then fresh specimens from Connecticut Valley Biological LM 48012DS, LM 48312DS, LM 48212DS, or preserved Ward's 63 W 0328, 63 W 0335, 63 W 0365)
- Coralline alga (Ward's 63 W 0368)
- forceps or latex gloves

Station 10:
- packages of edible seaweeds, dry and a portion soaked (from local natural food stores)

INVESTIGATIONS

If you have not gone on a field trip to see live specimens in the wild, your instructor may present a slide show to illustrate the diversity of fungal, lichen, and algal forms. Also, consult Figures 15.2 and 15.4 in this lab as well as figures in your textbook.

Station 1:
☐ Examine the slide of the ascomycete *Morchella*; it is a cross section of the cap region of the morel (see Figures 15.3 and 15.4); locate the densely packed asci lining the pits on the cap surface; a mature ascus will contain a column of clearly visible ascospores.

Station 2:
☐ Examine the slide of the basidiomycete *Coprinus*; scan the section until you locate the basidia (see Figure 15.3); use high power to see them clearly; if a basidium has already released its four spores, then there will be four points at the end of the basidium where the basidiospores used to be attached.

Station 3:
☐ Examine the crustose, foliose, and fruticose lichens under a dissecting microscope. Look for ascocarps on the specimens displayed.

15

Q9. On a lichen specimen with ascocarps:
What color are that lichen's ascocarps?

_____

What color is that lichen's upper surface?

_____

What color is that lichen's lower surface (if foliose)?_____
What other distinctive features (such as wrinkles, cracks, finger-like projections on the upper surface and root-like rhizoids underneath) can you see on that lichen under the microscope?

☐2. To see the location of the green photosynthetic cells in a lichen clearly, use a single-edge razor blade to cut across a fresh, hydrated foliose lichen; use forceps to hold the cut specimen such that you can examine the cut edge under a dissecting microscope; use the microscope's highest power to locate the green photosynthetic cells. Draw the cross section of the lichen below, labeling the photosynthetic cells.

Q10. Lichens reproduce vegetatively by means of small fragments that break off and are blown or washed to another location. A lichen, however, does not reproduce itself sexually; explain this.

Station 5:
☐After examining the prepared slides of *Morchella*, *Coprinus*, and lichen thallus at stations 1, 2, and 4, examine the two "unknowns" at this station; for each, determine whether it is a section of an ascomycete, basidiomycete, or lichen.

Q11. Unknown 1 is what kind of organism?

What diagnostic feature did you see?

Unknown 2 is what kind of organism?

What diagnostic feature did you see?

Station 4:
☐1. Examine the prepared slide of a lichen thallus (body) and locate the photosynthetic cells; in prepared slides, these photosynthetic cells are often stained reddish; but unfortunately, so are other parts of the lichen.

WORKSHEET 3: Fungi and Dispersal of Sexual Spores

Name of fungus	Location of sexual spores	How these sexual spores are dispersed

ascomycetes:

false morel

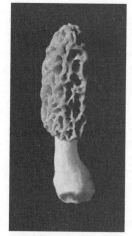

morel (*Morchella*)

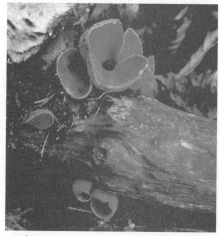

scarlet cup

basidiomycetes:

stinkhorn

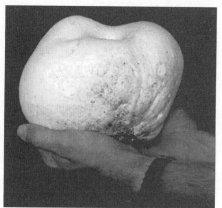

puffball

bird's nest fungus

Amanita

shaggy-mane (*Coprinus*)

shelf or bracket fungus

chantarelle

15

Figure 15.4 Diversity of ascomycete and basidiomycete fungi

173

Station 6:

☐ Examine the fruiting bodies displayed at this station. Consult Figure 15.4 to identify each specimen; reread the background section. For each specimen, figure out where the sexual spores are located and how they get dispersed. Note that mature asci typically explode, tossing spores into the air; basidiospores get dispersed by various methods such as by falling out of the basidiocarp, by the splashing action of rain, by insects, or by the rupturing of the basidiocarp. Record your information in Worksheet 3.

Station 7:

☐ Examine the lower surface of the various shelf fungi under the dissecting microscope.

Q12. Do all the shelf fungi on display, even the ones with lower surfaces that look smooth to the naked eye, have pores on their lower surfaces?

Q13. When a shelf fungus is present on a tree trunk, what does that tell you about the condition of the wood inside that tree trunk? Explain.

Q14. When a shelf fungus forms on a tree trunk, it always forms horizontally. Consider the microscopic structure and function of the shelf (reread the background section), consider what happens to the basidiospores upon release, and explain why it is most advantageous for the shelf to be horizontally rather than vertically oriented.

Q15. Sometimes you find a log on the forest floor, bearing a shelf fungus that is *vertically* oriented. Explain why this is so.

WORKSHEET 4: Algal Diversity

Unknown alga	Identified as	Its distinctive features
1		
2		
3		
4		
5		
6		
7		

Station 8:

☐ If the numbered algal cultures are not yet mounted and under microscopes ready for you, then make the wet mounts yourself (see Lab 2, "Microscopy and Plant Cells," for instructions). Examine the specimens and compare to Figure 15.2; identify each unknown algal specimen and record the distinctive feature it possesses on Worksheet 4.

Station 9:

☐ 1. These marine macroalgae are attached to rocks underwater or to intertidal rocks. Use Figure 15.2 and information in the background section to identify the numbered specimens on display.

Q16. Identify each of the numbered specimens:

1. =

2. =

3. =

4. =

☐ 2. Feel the texture of the macroalgae, using forceps or latex gloves if the specimens provided are preserved.

Q17. Which are flexible and rubbery?

How might this property provide a survival advantage in their environment?

Station 10:

☐ Seaweeds have a wide variety of uses. Certain seaweeds are the source of useful products such as carrageenan, alginates, and agar (see Part III of this lab). Other seaweeds are used as food—marketed as "sea vegetable" at times. The red alga *Porphyra*, known as nori, serves as the wrapper for sushi. This station displays some examples of seaweeds on sale for human consumption.

PART III:
SUPERMARKET SURVEY:
CARRAGEENAN, AGAR, ALGINATES
(homework—extended assignment)

OBJECTIVES

1. Become familiar with the diversity of grocery items containing algal ingredients.
2. Learn about the properties of agar, carrageenan, and alginates.
3. Learn about the diversity of uses for these algal extracts beyond grocery items.

BACKGROUND

Seaweeds, as well as extracts from seaweeds, are extensively used by humans. **Agar** and **carrageenan** are extracts from the cell walls of various red algae, and **alginates** are extracts from the cell walls of various brown algae. *Chondrus crispus*, the red alga you examined at station 9 in Part II of this lab, is one of the important sources of carrageenan. *Laminaria*, a brown alga also examined at station 9, is one of the sources of alginates.

In this homework assignment, you will investigate the diversity of grocery store items that contain these algal extracts. You will also research the properties of these extracts as well as the diversity of applications beyond the grocery store items that you found.

MATERIALS

- nearby grocery store
- Internet connection

INVESTIGATIONS

☐ 1. Go to a grocery store; find 20 different kinds of items that contain agar, carrageenan, or alginates in their list of ingredients. Record your findings in Worksheet 5.

15

☐2. Consult your textbook and other references to research the properties that these algal extracts impart to the products in which they occur. Also do an Internet search to find websites of well-established companies, such as FMC Corporation, that produce agar, carrageenan, or alginates; these websites give information on the diversity of products manufactured from these algal extracts. On Worksheet 6, record the properties of these algal extracts and the wide range of products that are made from these extracts. Cite below, the references and websites that you consulted.

LITERATURE CITED

Brockman, C. F. 2001. *Trees of North America: A Field Guide to the Major Native and Introduced Species North of Mexico*, rev. ed. (Golden Field Guides from St. Martin's Press). New York: St. Martin's Press.

Goward, T. 1992. Lichens. In: MacKinnon, A., Pojar, J., and Coupé, R., editors. *Plants of Northern British Columbia*. Edmonton, Alta.: Lone Pine Publishing. p. 319.

McKnight K. H., and V. B. McKnight. 1987. *A Field Guide to Mushrooms: North America*. (Peterson Field Guide series). Boston: Houghton Mifflin.

FMC Corporation. 2005. FMC Corporation "FMC BioPolymer," http://www.fmcbiopolymer.com/ (accessed February 19, 2006).

Product name and brand	Contains agar, carrageenan, or alginate?
1	
2	
3	
4	
5	
6	
7	
8	
9	
10	
11	
12	
13	
14	
15	
16	
17	
18	
19	
20	

15

Algal extract	Properties	Products
agar		
carrageenan		
alginates		

16 LAB

PLANT DIVERSITY:
Seedless Plants

I. Diversity in seedless plants (field trip)
II. Diversity in seedless plants (in class)

1. Know the distinguishing structural features of the various groups of seedless plants covered in this lab.
2. For each seedless group, be able to identify the sporophyte and know where the spores are produced.
3. Know the spore-dispersal mechanisms in the various groups of seedless plants.
4. For plants erroneously called "mosses," know the group of organisms to which each actually belongs.
5. Become familiar with the diversity within the ferns.
6. From observation of the seedless groups, deduce some evolutionary trends in the seedless plants.

NOTE: Today's lab is an example of **observational science** rather than experimental science. You will learn about the diversity of seedless plants by critically examining, comparing, and contrasting plant structures rather than by experimental testing of hypotheses. Much of our knowledge of plant structure and some plant processes comes from careful observations of many specimens over the years. Observations are also the foundation on which experimental science is built.

BACKGROUND

Humans have one body in their life cycle; all plants have two different bodies in sequence in their life cycle.

In *humans*, the sperm fertilizes the egg producing a zygote; this cell grows into a body that produces the single cells called gametes—the sperms or eggs—completing the life cycle. Thus, the human is a gamete-producing body.

In *plants* (see Figure 16.1), the sperm fertilizes the egg producing a zygote; this cell grows into a body which produces the single cells called **spores**, *which grow into a different body* that then produces the single cells called gametes—the sperms and/or eggs—completing the life cycle. Thus, plants have a spore-producing body (**sporophyte**) followed by a gamete-producing body (**gametophyte**) in their life cycle.

Genetically, the sperm and egg each carry one set of genetic information—one set of structures called chromosomes—a condition called **haploid** and symbolized as $1n$ or simply n. Fusion of the sperm with the egg in fertilization produces the zygote, which thus contains two sets of chromosomes, a condition called **diploid** and symbolized as $2n$. The plant zygote grows into a diploid sporophyte body that produces via reduction division (meiosis) the haploid spores; a spore grows into a haploid gametophyte body that produces haploid gametes—the eggs and/or sperms—completing the life cycle.

In contrast to seed plants, the **seedless plants** have the following features:

- *Spores* rather than seeds are the structures for dispersal to a new location. Seedless plants never produce seeds.
- Flagellated sperm cells are *released to swim through external water* to the egg. (In seed plants, the pollen grain contains the usually nonflagellated sperm cells and is transported by wind or pollinator to the vicinity of the egg.)
- The *zygote remains in place* on the gametophyte to grow into the sporophyte. (In seed plants, the zygote grows into an embryonic sporophyte that is contained along with nutrients within the seed that is then dispersed to a new location; upon seed germination, the sporophyte resumes growth.)
- The *gametophyte* colonizes a new location. (For seed plants, the sporophyte colonizes a new location).

16

The seedless plants include the liverworts, hornworts, and mosses, which are collectively referred to as the **bryophytes**. The bryophytes are **nonvascular**—they do not have specialized vascular tissue with lignin-containing cell walls for conducting fluids and nutrients through the body. There are also **vascular** seedless plants—the **lycophytes** (clubmosses, spikemosses, and quillworts) and the **pteridophytes** (whisk fern, horsetails, and ferns) (Greek: *bryon* = moss; *lykos* = wolf; *pteris* = fern; *phyton* = plant).

In this lab you will study liverworts, mosses, clubmosses, spikemosses, horsetails, and ferns. You will examine distinctive structural features and determine whether a structure is a gametophyte or a sporophyte, where the spores are produced, and how the spores are released and dispersed. Finally, you will deduce some evolutionary trends within the seedless plant group, based on your own observations.

PART I:
DIVERSITY IN SEEDLESS PLANTS
(field trip)

INVESTIGATIONS

NOTE: As always, when outdoors, be alert for such things as venomous animals and poison ivy.

☐ 1. Look in your textbook and familiarize yourself with the general appearance of the various groups of seedless plants. Your instructor will lead you to shady, damp campus areas, wooded edges, other natural areas and/or to the campus greenhouse; there you will search for seedless plants. Your instructor may point out the first few specimens and then ask you to find others. Search behind bushes on the north side of buildings; you may find liverworts alongside the building. If your campus has a wooded stream or ravine, you should find plenty of bryophytes and ferns. If there is a railroad embankment or eroded stream bank or ditch nearby, check for horsetails. In deserts, look for ferns in rock crevices of canyon walls, *Selaginella* on the ground, and occasional mosses under bushes or cacti (use caution).

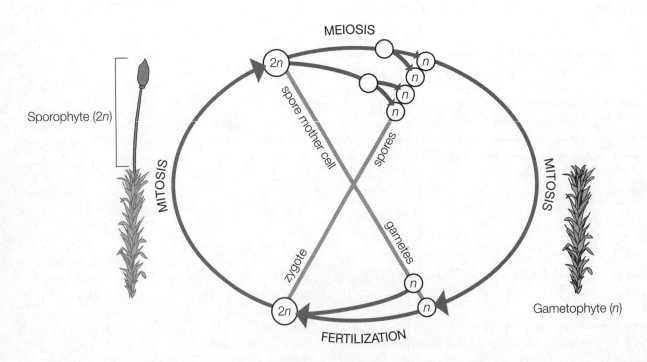

Figure 16.1 The plant life cycle (moss as the example). Note that every time you cross a line of the "X" in the diagram, there is a special cell that then undergoes a particular process to form something else.

☐2. In the greenhouse, look at the diversity of seedless plants in the potted plant collection; also look on the ground beneath the benches and on the soil surface of potted plants for seedless plant volunteers such as mosses, liverworts, and ferns. Interview your greenhouse manager, who is a gold mine of information and experience.

While on your trip, you may be making some of the observations described in Part II of this lab. Also, your instructor may collect specimens from the field, such as old moss sporophytes, that you will then examine more closely along with other specimens in class (Part II).

☐3. For each type of seedless plant found in the field, use the following Plant Log to note the location and environmental conditions in which it occurred.

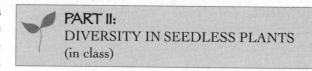

PART II:
DIVERSITY IN SEEDLESS PLANTS
(in class)

BACKGROUND

Here is some additional terminology useful in discussing seedless plants: The egg cell is located within a multicellular flask-shaped organ called an **archegonium** (Greek: *archegonos* = the first of a race). The archegonium-bearing structure is the **archegoniophore** (Greek: *pherein* = to bear). The flagellated sperm cells are produced in an organ called an **antheridium** (Greek: *antheros* = flowery; *–idion* is a diminutive ending). The antheridium-bearing structure is the **antheridiophore**. Similarly, a **sporangiophore** is a sporangium-bearing structure.

PLANT LOG: Field Trip Plants and Their Habitats

Plants Found	Location and Environmental Conditions There

16

MATERIALS

- Living materials listed below may be available from the campus greenhouse, or may be seen in the field.
- Compound microscopes are needed wherever a microscope slide is listed.

Station 1: dissecting scope, forceps, and paper towel
- leafy liverwort, living (*Porella* Carolina HT-15-6570)
- thallose liverwort, living (*Marchantia*: Carolina HT-15-6540, Ward's 86 W 4200)
  (living *Marchantia* with archegonia, most available in spring: Carolina HT-15-6546)
  (living *Marchantia* with antheridia, most available in spring: Carolina HT-15-6544)
- *Marchantia* preserved life cycle stages — (Carolina HT-22-3001) to be used if living ones not available
- safety glasses and gloves if preserved specimens are used (consult your supplier for safety information; Carolina and Ward's transfer their preserved specimens to a non-formalin holding solution for shipment, to minimize risk.)
- *Marchantia*, mature sporophyte, l.s. microscope slide (longitudinal section, Carolina HT-29-8776, Ward's 91 W 4048; median-longitudinal section, Ward's 91 W 4049)

Station 2: dissecting scope and dissecting needles
- various diverse mosses, living (from nature or Carolina HT-15-6710 (*Hypnum*), HT-15-6720 (*Mnium*), HT-15-6730 (*Polytrichum*))
- sporulating moss, living (Carolina HT-15-6695)
- old moss sporophyte capsules, from nature

Station 3: dissecting scope, compound scope, microscope slides, cover slips, dropper bottle of water, and forceps
- *Sphagnum*, living (Carolina HT-15-6740, Ward's 86 W 4400)
- dry coarse peat moss (greenhouse or local gardening center)

Station 4:
- Irish "moss" (living, Connecticut Valley Biological LM 48312DS [www.ctvalleybio.com/]; preserved, Ward's 63-W-0365)
- reindeer "moss" (Carolina HT-15-6440)
- club "moss" = *Lycopodium* (from station 5 plant)
- spike "moss" = *Selaginella* (from station 6 plant)
- Spanish "moss" (*Tillandsia usneoides* from craft store or gardening center)

Station 5:
- *Lycopodium*, living, running form, ideally with strobili (Carolina HT-15-6980)
- *Lycopodium* strobilus, l.s. microscope slide (Carolina HT-29-9836 , Ward's 91 W 4628)
- *Lycopodium lucidulum*, living (Carolina HT-15-6990)

Station 6:
- *Selaginella lepidophylla*, "resurrection plant" (Carolina HT-15-7016)
- *Selaginella*, living (Carolina HT-15-7016)
- *Selaginella* strobilus, l.s. microscope slide (Carolina HT-29-9878, Ward's 91 W 4641)

Station 7:
- leafy liverwort, living (*Porella* Carolina HT-15-6570)
- moss, *Dicranum*, living (Carolina HT-15-6698)
- *Lycopodium*, living, running form, ideally with strobili (portion of specimen in station 5)
- *Lycopodium lucidulum* (portion of specimen in station 5)
- *Selaginella*, living (portion of specimen Carolina HT-15-7016 in station 6)

Station 8:
- *Equisetum hyemale*, living, ideally with strobilus (Ward's 86 W 5300; Carolina HT-15-6950)
- *Equisetum* strobilus, microscope slide (l.s. Carolina HT-29-9740 , c.s./l.s. Ward's 91 W 4821)
- *Equisetum* spores and elaters (slide: Triarch 8-2EEEE; www.triarchmicroslides.com)

Station 9:
- an array of fern specimens of diverse leaf form, living or pressed specimens (tropical fern set, 5 living, Carolina HT-15-6842, Ward's 86 W 5555)
- various fern fronds, living or pressed specimens, showing various types of sori — such as: *Dryopteris, Polystichum, Asplenium, Adiantum, Pteridium, Onoclea* (from nature, herbarium, greenhouse, or local nursery)
- fern sorus with indusium, l.s., microscope slide (Carolina HT-29-9524, Ward's 91 W 4903)
- fern gametophyte with archegonia and antheridia, microscope slide (Carolina HT-29-9290, Ward's 91 W 4862)
- fern gametophyte with sporophyte, microscope slide (Carolina HT-29-9356, Ward's 91 W 4864)

INVESTIGATIONS

Work in pairs and visit each of the stations. At each station, read the corresponding material in this lab, and examine the specimens; then record your observations in the areas provided, answer the questions, and fill in the appropriate portions of Tables 16.1 through 16.3.

Station 1: Liverworts

☐1. Within the liverwort group, there are **leafy liverworts** and Y-branching ribbonlike **thallose liverworts**. The leafy specimen and the ribbonlike specimen here are the gamete-producing gametophytes.

☐2. In leafy liverworts, the "leaves" are arranged in two rows; some species have a third smaller row of "leaves" underneath. The delicate 1-cell-thick "leaves" of a leafy liverwort have no vein and thus are not true leaves; they often are lobed, with one lobe folded underneath the other. Look at a leafy liverwort under the microscope and *carefully draw* the leaf:

☐3. In leafy liverworts, the leafy gametophyte bears a sporophyte consisting of a thin, long stalk bearing one oval sporangium.

Q1. Does this specimen of leafy liverwort have a sporophyte? _____

☐4. *Marchantia* is a thallose liverwort. Look at the four preserved specimens of reproducing *Marchantia*. *NOTE*: Use safety glasses, gloves, and forceps to handle material preserved in liquid.

    a. The **gemma cup** contains packets of vegetative tissue called gemmae that are splashed out by raindrops; this is asexual reproduction since no egg or sperm are involved.

    b. The flat-topped umbrella-like **antheridiophore** has cavities in the top surface in which the flagellated sperm are produced.

    c. The palm-tree-shaped **archegoniophore** has microscopic flask-shaped archegonia underneath, each containing one egg. The stalk of the archegoniophore elongates after fertilization of eggs.

    d. After fertilization of the egg, the zygote stays in place and grows into a small **sporophyte**—positioned like a coconut on a palm tree.

Q2. What is the advantage of a *short* archegoniophore stalk before egg fertilization, and a *long* stalk at sporophyte maturity?

Q3. Is the living *Marchantia* specimen at this station in a vegetative, asexually reproducing, or in a sexually reproducing state? _____

_____

If sexually reproducing, are there archegoniophores, antheridiophores, or both? _____

_____

☐5. Look at the microscope slide of a longitudinal section of a *Marchantia* sporophyte. Locate the club-shaped sporophyte's very short stalk and the sporangium. In the sporangium, note the densely packed spore mass, throughout which are scattered elongated cells called **elaters** that have spiral wall thickenings. Once the sporangium has ruptured, the tightly packed spore mass is loosened by the embedded elaters twisting and turning with changes in humidity; this facilitates spore dispersal.
Draw some spores and elaters to show relative size and dimensions.

Station 2: True Mosses

☐1. Examine the diversity of moss shapes, from upright to creeping growth form, from narrow to broad "leaves." These are *not* true leaves because they contain no vascular tissue. Each of these leafy gamete-producing gametophytes bears "leaves" of all the same type, spirally positioned on the stem. "Leaves" of some mosses may have a nonvascular midrib extending part of the way across the leaf.

☐2. On the sporulating moss specimen, examine the sporophyte—consisting of the long stalk and cap-

16

183

sule at the end. The capsule shape differs with the species. The sporophyte grew up through the archegonium and may still have a hat-like remnant of the archegonium resting on top of the capsule. This archegonial remnant is called the **calyptra.**

Q4.   Do any of the capsules have a calyptra present? _____ If so, you can easily pull the calyptra off with forceps or your fingernails.

☐3. Examine a mature moss capsule under the dissecting scope. In true mosses when the spores are mature, the lid-like end of the capsule detaches, revealing a ring of teeth called the **peristome** in the capsule opening (Greek: *peri* = all around; *stoma* = mouth). In dry weather, these peristome teeth curl outward to permit spores to shake out; then in wet weather, the teeth curl back inward.

Q5.   Why would a system have evolved that promotes spore dispersal in *dry* weather? What is the advantage?

Station 3: *Sphagnum* Moss

☐1. Look at the soft, mop-like, leafy body of *Sphagnum*, a moss found in bogs and other wetland; it is the gamete-producing gametophyte. When a fistful of wet *Sphagnum* is squeezed, a surprisingly large amount of water runs out. *Sphagnum* can absorb a great deal of water and thus assists in flood control; it has also in the past been used as wound dressing and as diapers.

☐2. Examine a fresh *Sphagnum* leaf under the compound microscope. *Sphagnum* can hold large quantities of water due to its leaf structure; there are huge, long cells each rimmed by thin, small cells. At maturity each large cell forms ribs of wall thickenings for structural support; then the large cells die and openings form in their cell walls. These empty cells now serve as large reservoirs that can take in or release water. The thin, small cells between the large ones remain alive and photosynthetic. *Draw a Sphagnum leaf and label*: living cells, dead cells, ribs, openings.

☐3. Look for any dark, small, *round sporangia* on stalks of your *Sphagnum* gametophyte. If a sporangium is present, *examine it* under the dissecting microscope. A *Sphagnum* sporangium blasts off its lid and thereby also expels its spores; there is no peristome ring of teeth around the capsule opening, as there is in true mosses.

☐4. Examine a wet mount microscope slide of a bit of peat moss. Peat moss is mined from old bogs. The acidity of bogs inhibits decomposition.

Q6.   How does the structure of the peat moss compare to that of the fresh *Sphagnum* leaf?

Station 4: Misnamed "Mosses"

☐1. Examine these organisms that are misnamed. None of these are mosses of the bryophytes.
   – *Irish "moss"* is a soft, frilly, marine red alga. It is a source of carrageenan.
   – *Scale "moss"* is another name for leafy liverwort, described under station 1.
   – *Club "moss"* is actually a vascular seedless plant that you will examine in station 5.
   – *Spike "moss"* is also a vascular seedless plant that you will examine in station 6.
   – *Reindeer "moss"* is more correctly called reindeer *lichen*. Reindeer graze on these spongy lichens that are soft when wet but hard and brittle when dry. Lichen consists of two organisms growing together—the strands of a fungus twine around the cells of an alga or cyanobacterium.
   – *Spanish "moss"* is actually a flowering plant that hangs on other plants but is not parasitic.

☐2. On the Spanish moss, look for a clump of three long, thin, twisted brown strips that are the remnants of the dry fruit that has split open to release its fluffy seeds.

Station 5: Clubmosses—*Lycopodium*

☐1. The leafy clubmoss body is the sporophyte. The gametophyte is tiny and inconspicuous. Like other lycophytes, clubmosses are vascular, have simple single-veined leaves, and bear each sporangium on the upper surface of a leaf. The sporangium-bearing leaf is called a **sporophyll**. Sterile leaves, without sporangia, are **microphylls** (Greek: *phyllon* = leaf).

☐2. Examine the two kinds of *Lycopodium*. *Lycopodium lucidulum* bears its sporophylls, when present, in zones along the stem alternating with sterile zones. The more advanced *Lycopidium* give this group its common name of *club*moss due to the sporophylls being located in separate club-like cones called **strobili** (singular **strobilus**) held beyond the bulk of the sporophyte body (Latin: *strobilus* = pine cone).

Q7.   Why would it be advantageous to have sporophylls in strobili beyond the sporophyte rather than in zones along the leafy stem?

☐3. Look at the longitudinal section of a *Lycopodium* strobilus. An organism is called *homo*sporous if all the spores produced by the sporophyte are the *same* (the same size).

Q8.   Is *Lycopodium* homosporous?

Station 6: Spikemosses—*Selaginella*
☐1. Examine the two species of spikemosses and determine which one lives in the desert. *Selaginella lepidophylla* is well adapted to its southwestern U.S. desert environment; it is called resurrection plant because it becomes brown and curled up when dried out; it then uncurls, becomes green again, and resumes growth when rehydrated. The other *Selaginella* species is not a desert species and is much more delicate, lacking tough structural adaptations against desiccation.

☐2. The spikemosses, like advanced clubmosses, have their sporophylls in strobili. Look at the longitudinal section of a *Selaginella* strobilus under the microscope. An organism is called *hetero*sporous if the spores are of two *different* types—small microspores and large megaspores produced in separate sporangia.

Q9.   Based on your observations, is *Selaginella* homosporous or heterosporous?

☐3. Don't bother looking for the *Selaginella* gametophytes; they are microscopic. The spore undergoes cell division, subdividing itself to produce the gametophyte *within the original spore wall*! A microspore grows into a tiny gametophyte that produces sperm cells; a megaspore grows into a tiny gametophyte that produces egg cells.

Station 7: How to Distinguish Similar Mossy-Looking Bryophytes and Lycophytes
Review the background section and the information for stations 1, 2, 5, and 6; then read the material for this station and fill in Table 16.1.

Mossy-looking plants	Leaflike structures arranged spirally, *or* in how many rows down the stem?	Nonvascular or vascular?	Is this mossy structure 1*n* or 2*n*?	Does the mossy plant support a stalk with one sporangium, *or* bear sporangia on microphylls?	Are sporangia in strobili?	Homosporous *or* heterosporous?
leafy liverwort						
true moss						
primitive clubmoss						
advanced clubmoss						
Selaginella						

Table 16.1 Distinguishing the mossy-looking plants of different groups

16

Bryophytes:
– Mosslike plant is 1*n* gametophyte; its fertilized egg then grows into a sporophyte (stalk bearing one sporangium).
– Nonvascular; "leaves" lack vascular veins.
– All spores are alike.

Lycophytes:
– Mosslike plant is 2*n* sporophyte; its sporangia are on upper surface of microphyll leaves.
– Vascular; simple leaves (microphylls) have a single unbranched vein.
– Only some members have all their spores alike.

Station 8: Horsetails—*Equisetum*

☐1. Note the whorls of reduced leaves along the sporophyte shoots that grow from the underground rhizome stem of *Equisetum*, the horsetails. Examine the longitudinal section of the strobilus under the microscope. Look for a cone scale or **sporangiophore** that has been sliced through its middle, showing the umbrella-like sporangiophore structure bearing the sporangia hanging along its rim.

☐2. *Equisetum* gametophytes are hard to find in nature, being at most about 3 cm long.

☐3. Examine the horsetail spores on a prepared microscope slide if fresh ones are not available. At maturity when the stalks of the sporangiophores lengthen to open the strobilus, the sporangia split open. Then as the tightly packed spores dry out, the thick outer cell wall material of a spore unwraps as four strips, called **elaters**, that remain attached to the spore at one end. If you observe fresh horsetail spores under a dissecting microscope, the unwrapping movements of the elaters makes the green spores look like hibernating four-legged spiders that are waking up!

Q10. What advantage to the horsetails are these unwrapping elaters of the spores? (Hint: What did elaters do for liverworts?)

☐4. Today, horsetails are spindly herbaceous plants. However, check out "The Giant Horsetails" website at http://www.fiu.edu/~chusb001/giant_equisetum.html; these are impressive, although not the massive trees of the Carboniferous fossils! (If the address changes, then do a web search for giant horsetails.)

Station 9: Ferns

☐1. Look at the diversity of shapes of fern leaves, which are also called **fronds**. The fern leaves and thick rhizome stem from which they arise are parts of the sporophyte.

☐2. Look underneath the fern leaves; you will often find clumps or lines of brown granular material there. People have erroneously thrown ferns away, thinking they were sick or infested. These brown clumps are clumps of sporangia, *not* insect pests or insect eggs! The clumps are called **sori** (singular: **sorus**). The pattern of sori arrangement on a fern leaf is distinctive for the species.

Draw the sori arrangement on a leaflet of three different fern species at this station:

☐3. In some species the sorus is associated with a small membrane called an **indusium** that can have various shapes such as a kidney bean, an umbrella, or a lean-to that shelters the developing sporangia. Some ferns have a **false indusium** consisting of the leaf edge rolled under, sheltering the sporangia. Some species have no distinct sori, but rather entire leaflets of the leaf devoted to spore production. Other species have two kinds of leaves—one entirely specialized for spore production, the other entirely vegetative without any sporangia. Examine sori of different ferns under the dissecting microscope and *draw* at least two different indusium shapes that you observed.

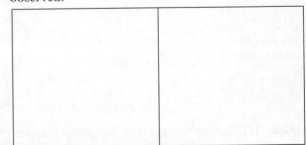

□4. Examine the microscope slide of a fern sorus. Note the cluster of lollipop-shaped sporangia. Sporangia of most ferns have a backbone-like line of prominent cells that extends most of the way around each sporangium. When spores are ready to be dispersed, the sporangium dries out, the belly of the sporangium slowly rips open, and the backbone gradually arches backward until it suddenly snaps forward, *catapulting* the spores into the air. *Draw* a sporangium below.

□5. Examine the fern gametophyte on the microscope slide. A fern spore grows into a small, thin gametophyte that is often heart-shaped. Near the notch are the archegonia that produce eggs, and closer to the point of the heart are antheridia that produce flagellated sperm. *Draw* the gametophyte and *label* the archegonia and antheridia.

□6. Examine the microscope slide of the fern gametophyte with emerging young sporophyte—the consequence of an egg in an archegonium having been fertilized by a sperm.

IN CONCLUSION

Fill in Tables 16.2 and 16.3. These summary tables are designed to help you review and organize the information that you have learned, and to help you gain an overview of the various groups of plants and their distinguishing characteristics.

Q11. Examine Table 16.3, which lists the seedless plants from the more primitive to the more advanced. What **evolutionary trends** can you see within the seedless plants—what features appear as you go down the table, which are not present in the more primitive groups? List them:

Plant	Spore-Dispersal Mechanisms
thallose liverwort	
true moss	
Sphagnum moss	
Equisetum	
Fern	

Table 16.2 Summary overview of the diversity of spore-dispersal mechanisms

16

Plant	Dominance of 1*n* gametophyte or 2*n* sporophyte?	Sporophyte branched or not?	Homosporous or heterosporous?	Number of sporangia on one sporophyte?	Lifespan of sporophyte: one season or years?	Vascular or not?	Approx. height of dominant body?
thallose liverwort							
true moss							
Sphagnum							
Lycopodium							
Selaginella							
Equisetum							
fern							

Table 16.3 Summary of characteristics of common seedless plants

EXTENDED ASSIGNMENTS

1. In reference books, look up the range and habitat of those seedless plants that you did not see during your field trip. Do they occur at all in your area of the country? If so, in what kind of habitat are they in your area?

2. In sexual reproduction of humans, <u>one</u> fertilization usually yields <u>one</u> offspring that disperses (eventually) and establishes itself elsewhere. Now consider sexual reproduction of seedless plants. For mosses and ferns, compare the impact of one fertilization event in terms of the potential number of offspring that can be produced to colonize other sites; a single fertilization yields a sporophyte bearing how many sporangia over what lifespan of the sporophyte? In conclusion, do the evolutionarily more advanced ferns have an advantage over mosses in some way other than size?

3. Service learning: The class can work together to produce a folder consisting of a campus map of seedless plants, accompanied by a descriptive paragraph for each species. The folder can be copied and made available to campus visitors.

LITERATURE CITED

Husby, C. 2003. "The Giant Horsetails," http://www.fiu.edu/~chusb001/giant_equisetum.html (accessed February 19, 2006). Miami, FL: Department of Biological Sciences, Florida International University.

LAB 17

PLANT DIVERSITY:
Seed Plants

I. Diversity in conifers
      Key to conifer genera
II. Diversity in flowering plants
Bibliography of field guides to trees (annotated)

OBJECTIVES

1. Learn some of the features that distinguish different genera of conifers.
2. Become familiar with the diversity of conifer leaves and of conifer cones.
3. Learn how to use a dichotomous key to identify a specimen.
4. Become familiar with the layout of various field guides and gain experience using them.
5. Become familiar with the diversity of flower structures in the flowering plants.
6. Become familiar with the diversity of leaf shapes and leaf arrangements in flowering plants.
7. Learn to distinguish between monocots and eudicots.
8. Learn to distinguish between primitive and advanced features in flowers.

NOTE: Today's lab is an example of **observational science** rather than experimental science. You will learn about the diversity of seed plants by critically examining, comparing, and contrasting plant structures rather than by experimental testing of hypotheses. Much of our knowledge of plant structure and some plant processes comes from careful observations of many specimens over the years. Observations are also the foundation on which experimental science is built.

BACKGROUND

Plant Classification

Plants are classified primarily on the basis of their reproductive parts. The plant classification system consists of a hierarchy of groups:

NOTE: Your instructor may have you focus only on certain parts of this lab.

kingdom
    phylum (plural is *phyla*); until recently, botanists
        used the name division rather than phylum
      class
        order
          family
            genus (plural is *genera*)
              species

Closely related species of organisms are grouped together into the same **genus**. A plant's binomial scientific name consists of two words, the genus and the specific epithet. There are various species of pine such as ponderosa pine (*Pinus ponderosa*), eastern white pine (*Pinus strobus*), and longleaf pine (*Pinus palustris*), all in the genus *Pinus*. Note that by convention the scientific name is italicized, and only the genus name is capitalized. The pines—along with other similar genera such as the spruces, firs, hemlocks, and so on—are all in the same **family**, Pinaceae. The cypress family Cupressaceae is similar enough to the Pinaceae to be together with it in the **order** Pinales. The yew family Taxaceae is different enough from all these to be put in its own order, Taxales. The Taxales and Pinales orders are similar enough to be in the same **class**, Pinopsida, within the **phylum** Coniferophyta (Greek: *phyton* = plant) within the **kingdom** Plantae.

Seed Plants

Seed plants disperse their offspring to new locations by means of seeds. Seed plants produce **pollen grains** that contain sperm cells, and they produce small oval or spherical **ovules,** each of which contains one or a few egg cells. When an egg cell in an ovule is fertilized by a sperm, that ovule then develops into a seed. A **seed** consists of a protective seed coat that encloses an embryonic offspring plant and nutritive tissue. The nutritive tissue will provide the embryo with energy and nutrients needed during seed germination. The

17

seed plants include two groups—the gymnosperms and the angiosperms.

In **gymnosperms**, the naked seeds develop from ovules that are not enclosed in a structure, but rather are exposed on a cone scale or a stalk (Greek: *gymnos* = naked; *sperma* = seed). The gymnosperms consist of four phyla: the conifers (pine, fir, spruce, cedar, etc.), the cycads, the ginkgo, and the unusual gnetophytes. You will see members of these phyla on campus, in the campus greenhouse, and/or pictured in your plant biology textbook.

The **angiosperms** are the flowering plants, all in one phylum. The flowers include female structures called pistils, inside of which are enclosed the ovules. Upon fertilization of the egg cells within the ovules, the pistils mature into fruit tissue, which contains the seeds (Greek: *angeion* = container; *sperma* = seed). The fruit tissue may be fleshy or dry and may split open at maturity to facilitate seed dispersal. See Lab 18, "Reproduction of Flowering Plants," for more details.

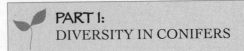

BACKGROUND

Conifers planted on campuses and in urban areas typically include native as well as nonnative species. If a conifer is native to California or China or Europe, it may well be missing or only briefly mentioned in a field guide to trees of the eastern United States. That is why it is useful to have a collection of various tree books and field guides available as references.

There is a great diversity in conifer structure, from creeping junipers to giant sequoias, from scalelike leaves 1 mm long to needles 46 cm (18 inches) long; from small, semi-fleshy, berrylike cones to woody, 66 cm (26 inch) long cones; from thin cone scales to hard, thick, woody ones armed with sharp spines.

Conifers are cone bearers (Latin: *conus* = cone; *ferre* = to bear). The male cones, also called pollen-bearing or staminate cones, often go unnoticed; they are small, soft, short-lived, and produce pollen grains (see Figure 17.2). The female cones, also called ovule-bearing or seed cones, are the ones that people notice (see Figure 17.1). The female cones of many conifers have two ovules on top of each cone scale (viewing the cone with the stalk end down). The ovules contain egg cells; when the eggs are fertilized, the ovules develop into seeds. The female cones are usually tough or woody in order to protect the developing seeds, some of which take more than a year to mature.

Note that although the ovule is initially naked on top of the cone scale, the cone scales close together tightly during seed development, effectively blocking entry to seed predators.

scales

male cones

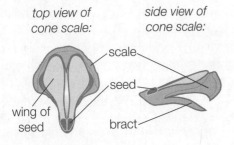

top view of cone scale: *side view of cone scale:*

scale
seed
wing of seed
bract

Figure 17.1 Ovule-bearing cone

Figure 17.2 Cluster of pollen-bearing cones

190

There are two small groups classified as conifers on the basis of various characteristics, although they do not have the typical female cone with tough cone scales. The families Taxaceae of the northern hemisphere and Podocarpaceae of the southern hemisphere have reduced modified "cones" consisting of a single ovule; during the ovule's maturation into a seed, tissue at its base proliferates to form a fleshy structure just beneath the seed (in *Podocarpus*) or to cover the seed partly (in *Taxus*) or wholly (in *Torreya* and *Cephalotaxus*).

I.A. CONIFERS IN THE FIELD (trip, and bad weather alternative)

Not all evergreens are pines. Not all conifers are evergreen. On this field trip, you will learn to use a key and field guides to identify conifers. By doing so, you will see some of the diversity of conifer leaves and reproductive structures. Depending on the season, your instructor may also have you examine flowering plant characteristics while in the field.

MATERIALS

- list of plants to be examined on the field trip (handout prepared by instructor)
- a variety of tree books and field guides (see the annotated bibliography at the end of this lab)
- Key to Conifer Genera Native to the United States and Canada (provided in this lab)
- campus grounds with conifers
- *Bad weather alternative:* a collection of slides or computer images to be projected

TRIP FORMAT

NOTE: As always, when outdoors, be alert for such things as venomous animals and poison ivy.

You may be instructed to work singly or in groups. Your instructor will indicate which one or more of the following active learning formats will be used in your field trip. You will be learning the identifying characteristics of various conifer species; you will also be researching information about the species—habitat, range, uses, etc.

Format 1. Each student researches and teaches about his or her selected species

Each student selects a species from a list provided and researches it in the reference books. Then, one of the following happens:

(a) The class *follows a map* that gives the *approximate* location of a specimen of each of the species on the plant list. At each approximate location, the student in charge of that species finds it, possibly with help from the group, and then teaches the class about it.

(b) Students are *sent out without a map* into an indicated area (a portion of campus) to find their species. Then the students return to *create a map and to decide on a circuit* for the class to walk. During the circuit walk, each student identifies and teaches about his or her species.

(c) The instructor leads the class but the *students call a halt* whenever a listed species is spotted; the student in charge of that species then identifies and teaches about it. The instructor may have to announce on occasion, "The class has just passed by a species from the list; go back and find it." Each student then determines whether it is his or her species that was missed.

Format 2. "Within sight of me is species X; where is it?"

All students consult reference books to learn the identifying characteristics and to locate the announced species; the first student to locate and identify correctly might get a reward. The books are consulted further for information about the species.

Format 3. All students work to identify and research an unknown specimen indicated in the field

The instructor leads the class to an unknown conifer, and all students determine the genus using the Key to Conifer Genera. Then the students check books for genus confirmation, species determination, and additional information. The first student to identify correctly might get a reward.

Format 4. If bad weather: projection of slides or computer images—armchair field trip

For each image projected by the instructor, either the whole class works to identify and research it using the key and books, or the student in charge of the species identifies it and teaches the class about it.

ASSIGNMENT

At the end of the trip, write a dichotomous key for only those species that you saw on your trip. A key is called dichotomous to highlight the fact that there are two choices to choose from at each step.

17

KEY TO CONIFER GENERA NATIVE TO THE UNITED STATES AND CANADA
(plus two alien genera often found on campuses)

Definitions

bract = a modified leaf; beneath each cone scale is a thin bract that in some species is longer than the cone scale and thus visible from the outside.

deciduous = describes a structure that is not permanent but falls off the plant; describes a species that drops all its leaves before the inhospitable season (before winter in temperate regions; before the dry season in deserts).

dioecious = describes a species where each plant has either all male reproductive structures or all female reproductive structures.

Directions

Examine the specimen; choose between description 1a and 1b; then follow its dotted line and go where directed. There, again choose one of the two descriptions, follow its dotted line, and so forth.

1a. Nonreproductive side shoots are densely covered
    by scalelike leaves 1–4mm long
    or by scalelike leaves with tapered points up to 12 mm long,
    or by spreading leaves, tapered, sharp-tipped, up to 15 mm long,
        in twos and threes,
    or a combination of these features .... go to 2

1b. Side shoots bear linear needles with edges parallel for most of the length
    arranged in bundles or singly on twig; needles of most species over 15 mm long .... go to 8

2a. Cones berrylike, semifleshy, bluish or reddish-brown;
    fleshy cone scales are grown together and do not separate at maturity.
    .... (the junipers and eastern red-cedar) **Juniperus**

2b. Cones leathery to woody; cone scales separate or detach at maturity .... go to 3

3a. Cones 2–3 cm, ball-shaped, fall apart at maturity; shoots grow vertically from branches.
    .... (pond cypress) **Taxodium**

3b. Cones woody or leathery; cone scales separate to release seeds but do not fall off at maturity .... go to 4

4a. Cones woody, pineapple-shaped, 5–8 cm long;
    scales separate at maturity but not widely .... (giant sequoia) **Sequoiadendron**

4b. Cones woody, less than 3.5 cm long .... go to 5

5a. Cones woody, spherical, resembling mini soccer balls before they open;
    ends of cone scales abruptly enlarged, not overlapping .... go to 6

5b. Cones woody, elongated; 6–12 cone scales overlapping until cone opens .... go to 7

6a. Cones 1.2–3.5 cm diameter, 6–12 scales .... (the cypresses) **Cupressus**

6b. Cones 0.6–1.2 cm diameter, 4–10 scales
    ..... (Atlantic white-cedar; others variously called cedars, cypresses, or false cypresses) **Chamaecyparis**

7a. Cones 1–1.2 cm long, held upright .... (northern white-cedar or arborvitae, western red-cedar) **Thuja**

7b. Cones oblong, 2–2.5 cm long, hang from tips; scales open wide at maturity
    .... (incense-cedar) **Calocedrus**

8a. Twigs older than 1 year have stubby, peg-like lateral shoots;
    the tip of each lateral shoot bears a dense, bouquet-like bunch of many needles .... go to 9

8b. Twigs do not have stubby pegs bearing dense bunch of needles .... go to 10

9a. Leaves deciduous, drop in autumn; cones 1–3.5 cm with thin scales................(larches, tamarack) **Larix**
9b. Leaves evergreen; cones large, up to 14 cm, solid barrel-shaped mass, upright on twig;
      mature cones disintegrate on tree................(true cedars from Mediterranean & Himalayas) **Cedrus**

10a. Needles in bundle of 2–5 (one species has 1 needle);
      bundle emerges from its own bud at end of tiny rudimentary side shoot,
      with bud scales then forming persistent or deciduous membranes surrounding base of bundle
      ....(the pines) **Pinus**
10b. Needles singly along the twig..go to 11

11a. Dioecious; no cones with cone scales; single seed has fleshy structure associated with it.............go to 12
11b. Not dioecious (both sexes on same plant); has cones with cone scales.......................go to 13

12a. Bell-like structure: red fleshy cup surrounds single seed................................(the yews) **Taxus**
12b. Plum-like structure: single seed completely surrounded by flesh,
      green streaked with purple.
      ....(nutmeg-trees—but not the true nutmeg, which is a flowering plant) **Torreya**

13a. Leaves deciduous, needles drop in autumn..go to 14
13b. Leaves not deciduous; evergreen..go to 15

14a. Leaves and side shoots alternate; cone ball-like, 2–3 cm diameter;
      scale outline on cone is roughly squarish diamond; cones fall apart at maturity.
      ....(pond cypress, bald cypress) **Taxodium**
14b. Leaves and side shoots opposite; cone ball-like, 1.6–2 cm diameter;
      scale outline on cone is 3–4 times as wide as high; cones remain intact,
      cone scales separate at maturity................................(dawn redwood, from China) **Metasequoia**

15a. Cones have scales overlapping, like shingles on a roof................................go to 16
15b. Cones have thick, blunt, woody ends; not overlapping................(coastal redwood) **Sequoia**

16a. Cones erect on twig; cones disintegrate on tree................................(the firs) **Abies**
16b. Cones hang from twig; cones remain intact, scales separate at maturity................go to 17

17a. Bract flat, three-pointed, looks like flattened hind end of a mouse (center point is like long thin tail);
      bract extends beyond cone scale, visible outside of cone................(the Douglas-firs) **Pseudotsuga**
17b. Bract not visible on the outside of the cone..go to 18

18a. Prominent ridges run lengthwise on twig surface; end of each ridge continues into a needle;
      when needle drops from old twig, a tiny woody peg remains................(the spruces) **Picea**
18b. No ridges on twig surface; base of leaf constricts abruptly to join a distinct thin leaf stalk.
      ....(the hemlocks—but not poison hemlock, which is a flowering plant!) **Tsuga**

NOTE: Certain common names such as "cedar" and "cypress" are used for members of several different genera. Therefore, botanists use scientific names in order to communicate precisely.

17

I.B. CONIFERS IN CLASS

BACKGROUND

Structurally, the cone is a compact shoot. In flowering plants, a shoot bears leaves and the lateral shoots grow out of the leaf axil, which is the angle formed between the leaf and the shoot. Similarly, each **female cone** of conifers is a compact shoot bearing highly modified, compact lateral shoots called **cone scales**, on top of which the ovules lie. Each cone scale is located in the axil of a thin, modified leaf called a **bract**. Depending on the species, the mature brown, papery bracts may be so long that they protrude from the cone, may be present but not protruding, or may not be apparent.

The **male cones** of conifers are simpler. The male cone is a shoot with modified leaves that bear pollen sacs on their undersides. There are no modified lateral shoots in male cones.

MATERIALS

- mature female cones of pine and Douglas-fir, one each per student pair (campus or dry preserved cone collection Carolina ER-22-3320)
- male pollen-bearing staminate cones (from campus or preserved Carolina ER-22-3330)
- forceps per student pair
- safety glasses and gloves per student if cones preserved in liquid are to be handled
- twigs from 10 different conifer species, labeled A, B, C, etc., and mature female cones from these same 10 conifer species, labeled 1, 2, 3, etc.; 1 set of twigs and cones per group of 4 students
- conifer cone collection, if twigs and cones are not available locally (dry cones Carolina ER-22-3320)
- a variety of tree books and field guides (see the annotated bibliography at the end of this lab)
- paper or tape to use in labeling

INVESTIGATIONS

NOTE: Use forceps, safety glasses, and gloves whenever handling material that is preserved in a liquid. (Suppliers identify the kind of liquid used in shipping their preserved specimens. Carolina and Ward's transfer their preserved specimens to a non-formalin holding solution for shipment, to minimize risk.)

Mature Female Cone

☐1. Work in pairs. Examine a **mature female cone** of pine and of Douglas-fir. You may need to pry apart the **cone scales** a bit to see the entire top and bottom surface of the scales. Locate the **bract** beneath the cone scale. The bract may have a rounded end with or without a bristle, or it may be three-pointed with the central point a long bristle, making the bract look like the flattened hind end of a mouse.

Q1. Does the pine or the Douglas-fir have bracts protruding from the cone? _____

Q2. Draw the shape of the bract, if present.
pine: *Douglas-fir:*

☐2. At the base of the upper surface of the cone scale, find two **seeds** or two depressions where the seeds used to be before they were dispersed.

☐3. Many species of pine produce **winged** seeds— each seed has a flat, dry extension of the seed coat called a **wing**. If all the seeds have been dispersed, you can still deduce from the cone whether the seeds had wings. If the two seeds had wings, there will be two wing-shaped marks of a different shade of brown on the upper surface of the cone scale where the wings had lain.

Q3. Circle the species that has or had winged seeds: *pine   Douglas-fir*

Q4. How would you expect winged pine seeds to be dispersed?

How would nonwinged pine seeds such as those of the pinyon (or piñon) pine be dispersed? (The pine nuts on your salad are from the pinyon pine.)

Q5. In what part of the pine cone did you find *fully developed* seeds, or depressions where such full-sized seeds had been?
☐ the entire cone
☐ the cone's base
☐ the cone's middle
☐ the cone's tip
Number of seeds in your pine cone: _____
A tree with 100 such cones would yield:
_____ seeds!

Q6. Why would a tree produce so many seeds when the tree needs only one seed to replace itself? What will happen to all these seeds in the ecosystem?

Q8. Why have male cones not evolved to be woody? Why would such an energy investment not be of selective advantage?

Male Cone

☐1. Examine the male cones. On the lower surface of the cone's many small, leaflike structures, locate the pollen sacs—they may already have broken open and shed their pollen.

Q7. Figure 17.2 shows the numerous male cones clustered on a single conifer shoot. Consider the huge quantity of pollen that a single tree will release (to the detriment of people who are allergic to it). Why is it advantageous for the conifer to spend so much energy making this large amount of pollen? (Hint: Consider the method of pollination.)

Match It!

☐1. Each group of four students has a collection of different conifer twigs labeled A, B, C, etc., and a collection of their female cones labeled 1, 2, 3, etc. Use the Key to Confer Genera and/or tree reference books to identify and match the cones and twigs. Label each matched twig and cone set with its name. Record your results in the table provided. (If twigs and cones are not available locally, then use the purchased cone collections provided, one collection per student group. Use books to identify the preserved cones; record leaf characteristics for each species.)

☐2. When student groups are finished matching, each group switches to another table and checks that group's identifications for accuracy.

MATCH IT!			
Twig	= Cone #	= Name of species or genus	Distinctive characteristics
twig A			
twig B			
twig C			
twig D			
twig E			
twig F			
twig G			
twig H			
twig I			
twig J			

17

BACKGROUND

Flowering plants are classified primarily on the basis of their flower structure. Among the flowering plants, flower structure is extremely diverse. *The basic plan is the same in all flowers:* Moving inward, there are the sepals, then petals, then stamens, then one or more pistils.

Study Figures 17.3 and 17.4. *Some of these flower parts may be unusually shaped, reduced, missing or fused together.* Flower parts of the same kind can be fused to each other—such as the petals of petunia. A flower part can be fused to a different kind of flower part—petunia filaments are fused with the petals. Grass flowers are reduced, with petals missing (see Figure 17.5a). A daisy is actually a cluster of many flowers—the tubular disc flowers in the middle surrounded by the strap-shaped ray flowers (see Figure 17.5b).

PISTIL (named for its shape; Latin: *pistillum* = pestel)
= the female structure (composed of stigma, style, ovary)
• Tulips have one pistil per flower; strawberries have many pistils per flower.

   Stigma: (Greek: *stigma* = a mark, spot)
   • The receptive region that secretes a sugary solution in which a pollen grain can germinate.

   Style: (Greek: *stylos* = column)
   • Connects the stigma to the ovary.
   • Pollen tube grows through it to get to the ovules in the ovary.

   Ovary: (Latin: *ovum* = egg)
   • Contains one or more round ovules.

      Ovule: (diminutive of *ovum*)
      • One egg cell forms inside the ovule tissue.
      • After the sperm travels down the pollen tube and fertilizes the egg in the ovule, the ovule matures into a seed.

STAMEN (Latin: *stamen* = thread)
= the male structure (composed of anther and filament)
• Lilacs have two stamens per flower, tulips have six, and strawberries have many stamens per flower.

   Anther (Greek: *anthos* = flower)
   • Also called "pollen sac."
   • Splits open to release pollen.
   • Anthers in geraniums fall off easily once pollen is shed.

   Filament (Latin: *filare* = to spin)
   • Thin stalk supporting the anther.

PERIANTH (Greek: *peri* = about, *anthos* = flower)
= all of the sepals and petals in a flower

   Petals (Greek: *petalon* = petal)
   • Enclose and protect the reproductive parts.
   • Color attracts pollinators.
   • May be reduced or absent in wind-pollinated flower.

   Sepals (Latin: *sepalum* = covering)
   • Enclose and protect the flower.
   • May be colorful (lily) to attract pollinators, or green (rose).

Receptacle: The fleshy tip of the flower stalk to which the flower parts are attached.

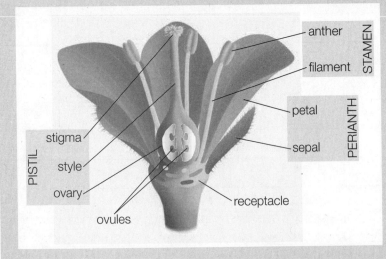

Figure 17.3 Parts of a flower

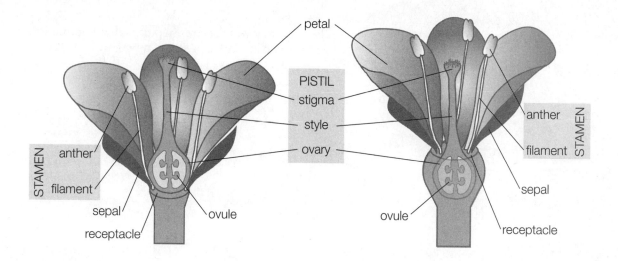

(a) Flower with **superior** ovary: Ovary is *above* the point of attachment of the other flower parts (tomato, peach, cherry, orange, etc.).

(b) Flower with **inferior** ovary: Ovary is *beneath* the point of attachment of the other flower parts (squash, blueberry, cranberry, apple, etc.).

Figure 17.4 Superior versus inferior ovary

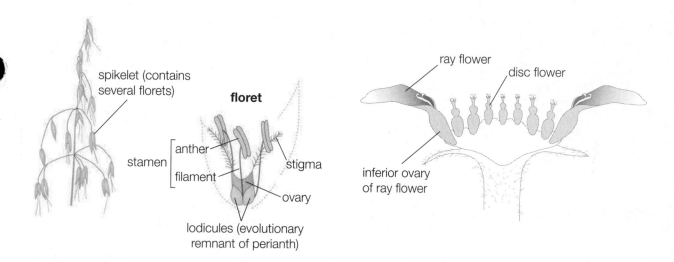

(a) Grass: A wind-pollinated flower that has evolved long dangly stamens, hairy stigmas, lots of pollen, and has lost the showy petals.

(b) Daisy: A head of many flowers crowded together; strap-shaped ray flowers on the outside; tubular disc flowers in the middle.

Figure 17.5 Modified flowers

Secondary to flower structure, the form and arrangement of leaves also enter into the classification and identification of plants. Leaf form is also extremely diverse. Figure 17.6 illustrates some basic leaf forms. To determine whether a leaf has more than one blade, look for the location of the axillary bud. Each leaf has an axillary bud in its leaf axil, which is formed by the leaf stalk (petiole) and twig. Compound leaves do not have buds at the base of their leaflets. Note that some species have small buds. In the sycamore, the petiole base completely surrounds and encloses the bud. Once you have determined what constitutes a leaf on your plant, you can then determine its leaf arrangement (see Figure 17.7).

17

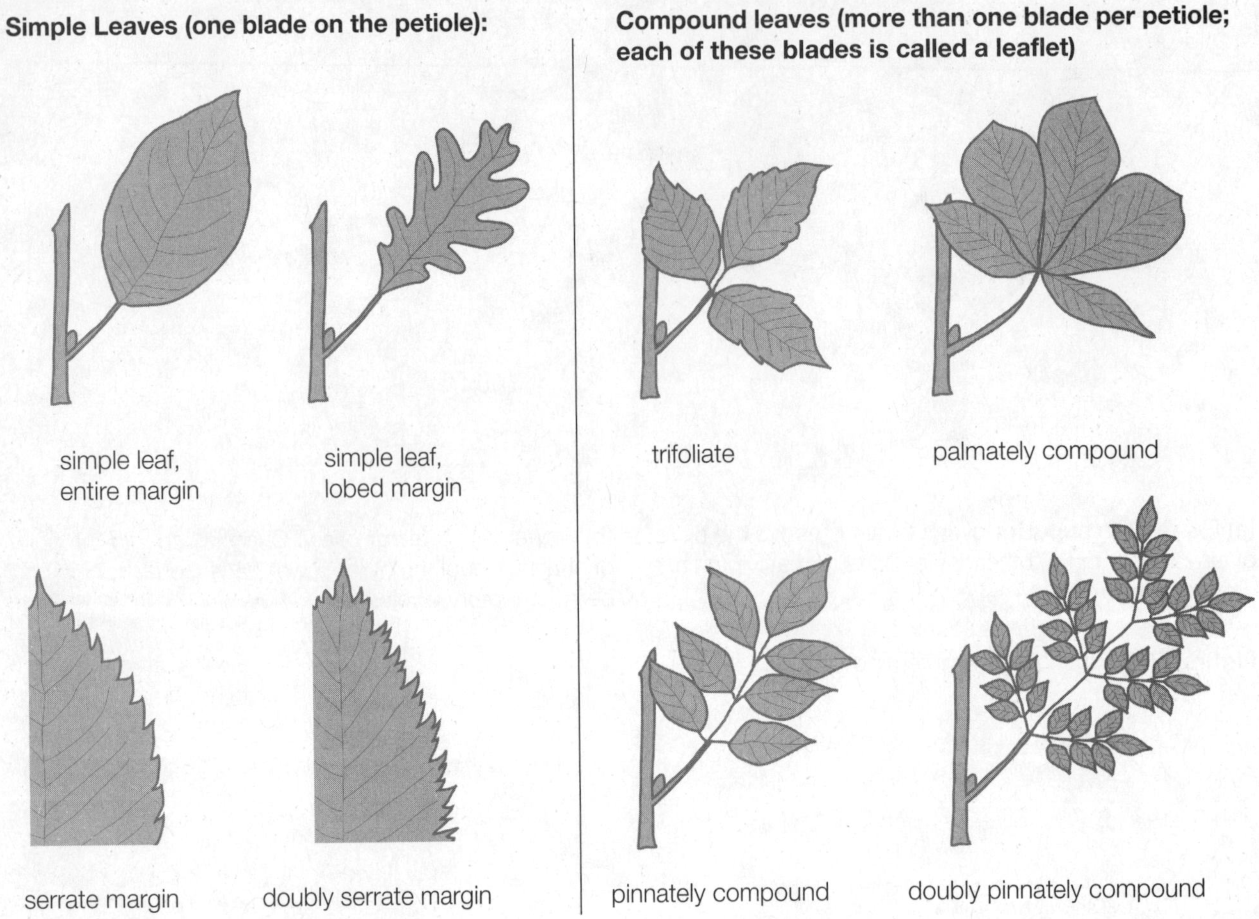

Simple Leaves (one blade on the petiole):

Compound leaves (more than one blade per petiole; each of these blades is called a leaflet)

simple leaf,
entire margin

simple leaf,
lobed margin

trifoliate

palmately compound

serrate margin

doubly serrate margin

pinnately compound

doubly pinnately compound

Figure 17.6 Basic leaf types

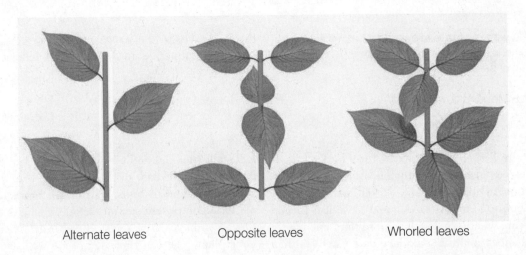

Alternate leaves

Opposite leaves

Whorled leaves

Figure 17.7 Leaf arrangements

MATERIALS

- flowering plants on campus or potted plants in/from the greenhouse or cut flowers; some possibilities are: *Magnolia*, *Alstroemeria*, false sea onion, *Zebrina*, lily, asparagus-fern, daffodil, tulip, dogwood, crabgrass, chickweed, dandelion, daisy, sunflower, petunia, *Nicotiana*, *Geranium*, *Begonia*, *Hibiscus*, pea, snapdragon, and *Impatiens*. Reduced flowers: the arums, *Peperomia*, and poinsettia. Unusual floral structures: *Bougainvillea*, passion flower, bird-of-paradise, milkweed, and orchid.
- dissecting microscopes and dissecting needles if flowers provided are small
- reference book with excellent line drawings of floral structures for each plant family, in case of questions: Lawrence, G. H. M. 1951. *Taxonomy of Vascular Plants*. New York: Macmillan.

INVESTIGATIONS

☐1. Carefully examine each of the flowers provided. Refer to the figures in this lab as you analyze each flower. (Structure of unusual flowers available in class will be explained by your instructor.) *Record* your observations on Data Sheet 1.

☐2. The vast majority of the flowering plants fall into one of two large groups:

The **monocots**:
- flower parts in multiples of 3
  (example: 3 sepals, 3 petals, 6 stamens, etc.)
- leaves with parallel veins
- one cotyledon (seed leaf) attached to embryo in seed

The **eudicots**:
- flower parts in multiples of 4 or 5
  (example: 4 sepals, 4 petals, 4 stamens, etc.; example: 5 sepals, 5 petals, 10 stamens, etc.)
- leaves are net-veined
- two cotyledons (seed leaves) attached to embryo in seed

The magnolia and the water-lily families are two evolutionary lines that diverged early and are neither monocot nor eudicot.

Record on your data sheet whether each plant appears to be monocot, eudicot, or neither.

☐3. Leafy shoots often have leaves spirally arranged along their length. A flower is an evolutionarily modified shoot. Flower parts are modified leaves. A **simple pistil** or **carpel** is a modified folded leaf containing ovules. A pea flower has a simple pistil that matures into the pea pod. If you split open a pea pod and flatten it, you see how leaflike it is. A **compound pistil** is the evolutionary result of several carpels fused lengthwise. A cherry tomato flower has a compound pistil with two carpels. This compound pistil matures into the cherry tomato, which shows the two carpels as two seed-filled cavities separated by a crosswall.

Table 17.1 shows several evolutionary trends present in flowering plants. Use the information in this table to evaluate the features of each of your flowers and determine whether there is a preponderance of primitive or advanced features. *Record* your conclusions on Data Sheet 1.

Primitive flower = magnolia, on the front cover	Advanced flower
Radial symmetry: ❀	Bilateral symmetry: symmetrical about one axis only
Petals and sepals look the same	Petals are distinct from sepals
Many flower parts of an indefinite number, spirally arranged on elongated receptacle	Few flower parts of a precise number, arranged in whorls on shortened receptacle
Flower parts still somewhat leaflike (i.e., long pollen sacs along strap-shaped stamen)	Flower parts less leaflike (i.e., distinct anther at end of thin filament)
Flower parts separate (i.e., simple pistils; separate petals and stamens)	Flower parts fused (i.e., compound pistil; petals fused into tube)
Superior ovary	Inferior ovary
Sepals, petals, stamens and pistils present in flower	Loss of one or more of the four kinds of flower parts (i.e., male flowers, female flowers)

Table 17.1 Evolutionary trends in flowering plants

17

Name of plant:						
leaf arrangement						
leaf type						
leaf margin						
# of sepals						
# of petals						
are petals fused to each other?						
# of stamens						
are stamens fused to each other?						
are stamens fused to other parts? which?						
superior or inferior ovary?						
# of pistils?						
monocot, eudicot, or neither?						
overall: primitive or advanced?						

EXTENDED ASSIGNMENT

The class can work together to produce a folder consisting of a campus conifer map accompanied by a descriptive paragraph for each species. The folder can be copied and made available to campus visitors. The map may cover only part of campus or may cover an adjacent neighborhood or nearby park.

BIBLIOGRAPHY OF FIELD GUIDES TO TREES (annotated)

NOTE: Nonnative species growing in urban areas (campuses) may not be covered in some of the books.

The Whole United States

Brockman, C. F. 2001. *Trees of North America: A Field Guide to the Major Native and Introduced Species North of Mexico*, rev. ed. (Golden Field Guides from St. Martin's Press). New York: St. Martin's Press. (280 pp.)

A really nice, compact book covering the whole United States and Canada—eastern, western and southern tree species. The figures are color drawings; brief description of each species accompanied by range map. Includes yuccas and palms.

Northeastern United States

Graves, A. H. 1992. *Illustrated Guide to Trees and Shrubs: A Handbook of the Woody Plants of the Northeastern United States and Adjacent Canada*, rev. ed. New York: Dover. (271 pp.)

This is a Dover reprint of the 1956 revised edition book that was published by Harper & Brothers, New York. This book also includes brief descriptions of some introduced species. Excellent detailed pen-and-ink drawings of leaves, twigs, and buds; good dichotomous "summer key" and "winter key" in front. This older book has some pieces of information very useful in identification that the other field guides do not have.

Harlow, W. M. 1957. *Trees of the Eastern and Central United States and Canada*. New York: Dover. (288 pp.)

This is a Dover reprint of a book originally published in 1942 by McGraw-Hill Book Company; tons of information in a small book. For each tree species, there are black-and-white photos of bark, twig, leaf, and fruit or cone; the text often includes information on how the pioneers and native Americans used the tree. A great little book. Another oldie but goodie.

Petrides, G. A. 1986. A *Field Guide to Trees and Shrubs: Northeastern and North-Central United States and South-eastern and South-Central Canada*, 2nd ed. (Peterson Field Guide series). Boston: Houghton Mifflin. (428 pp.)

This book is a must in the northeastern United States, since it *includes the shrubs and woody vines*, which are not included in other "tree" field guides and for which no separate field guide exists. The figures consist of clear, simplified line drawings of the leaves or needles. Does not include the Deep South.

Eastern United States

Little, E. L. 1980. *National Audubon Society Field Guide to Trees: Eastern Region* (Audubon Society field guide series). New York: Knopf. (714 pp.)

The figures consist of color photographs of bark and leafy twigs of each species, then a section of photos of tree flowers, then a section of photos of tree fruits, and finally a section on colorful autumn leaves. Line drawings of range maps, winter silhouettes, and fruits accompany many of the text descriptions.

Petrides, G. A. 1998. A *Field Guide to Eastern Trees: Eastern United States and Canada, Including the Midwest*. 1st ed., expanded (Peterson Field Guide series). New York, NY; Houghton Mifflin Company. (424 pp.)

Note that this Peterson Field Guide of eastern trees does *not* include the shrubs and vines but does include the trees of the Deep South; it has color drawings of the leaves and needles.

Western United States

Little, E. L. 1980. *National Audubon Society Field Guide to Trees: Western Region* (Audubon Society field guide series). New York: Knopf. (639 pp.)

The figures consist of color photographs of bark and leafy twigs of each species, then a section of photos of tree flowers, and a section of photos of tree fruits. Yuccas, palms, and large cacti are included. Line drawings of range maps, winter silhouettes, and fruits accompany many of the text descriptions.

Petrides, G. A. 1998. A *Field Guide to Western Trees: Western United States and Canada*, 1st ed., expanded (Peterson Field Guide series). New York: Houghton Mifflin. (428 pp.)

This field guide has plates of colored drawings; included are conifers, broadleaf trees, palms, large cacti, and yuccas.

Some Selected Trees

Coombes, A. J. 2002. *Trees* (Smithsonian handbooks). London: Dorling Kindersley. (320 pp.)

This book is a selection of trees native to the temperate regions of the world, both northern and southern hemisphere. It includes most of the tree species planted in urban areas. It includes various cultivated varieties in the nursery trade, which makes this book useful in identifying an unusual tree on campus. The figures consist of photographs of plant parts on a white background. This is a good book to have as a reference; it is, however, not as easy for beginners to use as one of the standard field guides because species are arranged by plant family.

17

Lanzara, P., and M. Pizzetti. 1978. *Simon & Shuster's Guide to Trees: A Field Guide to Conifers, Palms, Broadleafs, Fruits, Flowering Trees, and Trees of Economic Importance* (Simon & Schuster nature guide series; U.S. editor: Stanley Schuler). New York: Simon & Schuster. (317 pp.)

Only a selection of trees are covered. A useful reference for color photos of cloves, cashew fruits, almond fruits, and so forth as they appear on the plants. This book has an introductory chapter on tree biology, followed by species descriptions divided into six sections as indicated in the title. Each species has a half-page photo and half-page text with line drawings in the margin.

18

REPRODUCTION OF FLOWERING PLANTS:
Flowers to Fruits and Seeds

I. Flowers
II. Fruit and seed structure
III. So, what good is knowing stuff about flowers and fruits?

NOTE: Today's lab is an example of **observational science** rather than experimental science. You will learn about flower, fruit, and seed structure as well as about fruit formation through careful observation of specimens rather than through experimental testing of hypotheses. Much of our knowledge of plant structure and some plant processes comes from careful observations of many specimens over the years.

The Need of Being Versed in Country Things

The house had gone to bring again
To the midnight sky a sunset glow.
Now the chimney was all of the house that stood,
Like a pistil after the petals go.

The barn opposed across the way,
That would have joined the house in flame
Had it been the will of the wind, was left
To bear forsaken the place's name.

No more it opened with all one end
For teams that came by the stony road
To drum on the floor with scurrying hoofs
And brush the mow with the summer load.

The birds that came to it through the air
At broken windows flew out and in,
Their murmur more like the sigh we sigh
From too much dwelling on what has been.

Yet for them the lilac renewed its leaf,
And the aged elm, though touched with fire;
And the dry pump flung up an awkward arm;
And the fence post carried a strand of wire.

For them there was really nothing sad.
But though they rejoiced in the nest they kept,
One had to be versed in country things
Not to believe the phoebes wept.

—Robert Frost

OBJECTIVES

1. Learn the structures of a flower and how they facilitate floral functions.
2. Become familiar with the ways that flower structure varies from species to species.
3. Learn the floral origin of the various structures of a fruit.
4. Understand the fruit structure of a variety of familiar fruits, "vegetables," and nuts.

BACKGROUND

Flowers and Fruits in Perspective

Flowers are the reproductive structures of flowering plants. Flowers can range from being very colorful and conspicuous, such as a rose or orchid, to being very simple, reduced, and inconspicuous, such as those of grasses, oaks, and elms. The *function* of a flower is to ultimately produce a fruit that contains *seeds*—which then propagate the species. Every fruit has developed from a flower—an acorn came from a flower.

The **angiosperms** are the flowering plants; the seeds of flowering plants are enclosed in fruit tissue, which may be fleshy or dry.

The **gymnosperms**, such as pine, fir, spruce, cedar, and ginkgo, are also seed plants but not flowering plants; they have "naked seeds," as the name indicates. Their seeds are *not* enclosed within fruit tissue but rather are exposed, typically lying on top of a cone scale. In ginkgo, the naked seed dangling at the end of a stalk can be mistaken for a fruit because its *seed coat* is fleshy at maturity.

Knowledge of plant reproduction is critical to successful crop production. Certain plant structures such as tomatoes, eggplant, squash, pumpkin, peppers, pea pods, and string beans are **fruits**, botanically speaking. Each came from a flower and contains seeds.

18

The culinary designation of **vegetable** is based on the use of the plant part—it is eaten as part of the main course in a meal. Vegetables are actually various plant parts. In addition to those vegetables just listed that are fruits, there are vegetables that are botanically speaking **vegetative**, or nonreproductive parts: leaf stalks (celery), leaf blades (spinach), lateral buds (brussels sprouts), a young shoot (asparagus), a massive flowering structure in bud stage (broccoli), a root (sweet potato), an underground storage stem (white potato), and an entire, unelongated, aboveground plant (cabbage).

PART I:
FLOWERS

How Do You Tell Them Apart? What's That Stuff in the Middle?

Study Figure 18.1 to review the parts of a flower. Petals and stamens are often lost when a flower ages. The remaining pistil with its wide ovary does indeed look like an old fireplace chimney remaining after a house fire, as described in Robert Frost's poem.

PISTIL (named for its shape; Latin: *pistillum* = pestel)
= the female structure (composed of stigma, style, ovary)
- Tulips have one pistil per flower; strawberries have many pistils per flower.
    Stigma: (Greek: *stigma* = a mark, spot)
    - The receptive region that secretes a sugary solution in which a pollen grain can germinate.
    Style: (Greek: *stylos* = column)
    - Connects the stigma to the ovary.
    - Pollen tube grows through it to get to the ovules in the ovary.
    Ovary: (Latin: *ovum* = egg)
    - Contains one or more round ovules.
        Ovule: (diminutive of *ovum*)
        - One egg cell forms inside the ovule tissue.
        - After the sperm travels down the pollen tube and fertilizes the egg in the ovule, the ovule matures into a seed.

STAMEN (Latin: *stamen* = thread)
= the male structure (composed of anther and filament)
- Lilacs have two stamens per flower, tulips have six, and strawberries have many stamens per flower.
    Anther (Greek: *anthos* = flower)
    - Also called "pollen sac."
    - Splits open to release pollen.
    - Anthers in geraniums fall off easily once pollen is shed.
    Filament (Latin: *filare* = to spin)
    - Thin stalk supporting the anther.

PERIANTH (Greek: *peri* = about, *anthos* = flower)
= all of the sepals and petals in a flower
    Petals (Greek: *petalon* = petal)
    - Enclose and protect the reproductive parts.
    - Color attracts pollinators.
    - May be reduced or absent in wind-pollinated flower.
    Sepals (Latin: *sepalum* = covering)
    - Enclose and protect the flower.
    - May be colorful (lily) to attract pollinators, or green (rose).

Receptacle: The fleshy tip of the flower stalk to which the flower parts are attached.

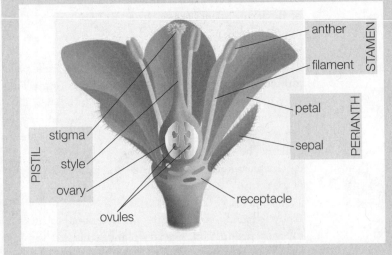

Figure 18.1 Parts of a flower

Pistil versus Carpel

A **simple pistil**, as in the pea flower, evolved from one folded leaf. Open up a pea pod (mature pistil), flatten it out, and you will see how leaflike it still is.

A **compound pistil**, as in the tomato or green pepper flower, evolved from several folded leaves that have over evolutionary time fused into one unit. Each *subunit* of a *compound pistil* is called a *carpel* (Greek: *karpos* = fruit). In a cherry-tomato fruit (mature ovary), the two carpels are still distinctly delineated by the cross-wall; in a green pepper, the cross-walls are only partially present. A simple pistil can also be called a carpel.

Note that you can determine whether a pistil is compound by taking a cross section of the ovary, or even by looking carefully at the outside of the ovary wall for indentations marking the boundaries of carpels. In begonias, the three carpels of the pistil are fused in the ovary but the three styles are distinct, unfused; in lilies, the three carpels of the pistil are fused their entire lengths, with the stigma showing three lobes.

Variation in Flowers

There is tremendous diversity in flower structure in the plant kingdom. The shape, size, color, position, and number of each kind of flower part present in a flower are distinctive for each species. For example, the flower of a species may have a superior or an inferior ovary (see Figure 18.2), or even an intermediate form.

Within a species, the *flower* structure is quite constant and is the *least variable part* of the plant—not much affected by growing conditions. A plant's flowers have evolved to be structurally well suited for a particular pollination method. In contrast, vegetative characteristics such as the size, shape, and number of leaves—as well as the height and degree of branching of the whole plant—can vary a great deal with climatic conditions. Flower structure is the basis of the current classification system of angiosperms.

Pollination versus Fertilization

The terms *pollination* and *fertilization* are often imprecisely used. **Pollination** refers to the deposition of pollen onto the stigma of a flower. This can be done by the wind or by various animals that have been attracted to the flowers—usually to feed on nectar or pollen.

Once the flower is pollinated, the pollen grain **germinates** in the sugary liquid on the stigma and grows a **pollen tube** down through the style into the ovary and into an ovule. See Figure 18.2a. A sperm cell travels down the pollen tube, and fuses with or **fertilizes** the egg cell in the ovule. How the pollen tube finds its way to the ovule and the egg is not fully understood; the mechanism of pollen tube guidance is an active field of research.

The fertilized egg, called a **zygote**, then develops into the embryonic plant that is in the seed; the seed is a matured ovule.

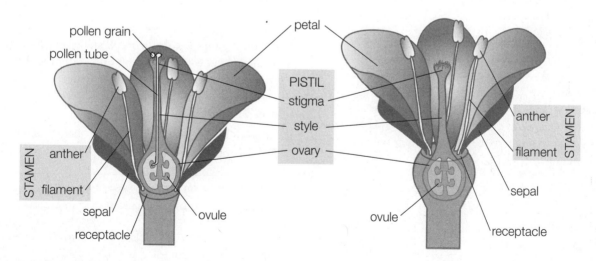

(a) Flower with **superior** ovary: Ovary is *above* the point of attachment of the other flower parts (tomato, peach, cherry, orange, etc.). Pollen tube is shown.

(b) Flower with **inferior** ovary: Ovary is *beneath* the point of attachment of the other flower parts (squash, blueberry, cranberry, apple, etc.).

Figure 18.2 Superior versus inferior ovary

18

Q1. Grasses are typical of open, windy environments. Grasses flowers have evolved long, dangling stamens, hairy stigmas, and very reduced or absent petals and sepals. Why would such floral features be advantageous to a grass?

Q2. What could be the adaptive advantage of having an inferior ovary? (Hint: herbivorous insect)

I.A. FLOWER STRUCTURE IN GENERAL

MATERIALS

- large flower (1 per student):
  - Group A—petunia, flowering tobacco (*Nicotiana*), tulip, geranium, Asian lily, or *Kalanchoë*
  - Group B—*Alstroemeria*, daffodil, or *Fuchsia*
- broccoli with large buds
- dissecting microscope (1 per student)
- dissecting needles (2 per student)

INVESTIGATIONS

☐1. At each lab table, half the students will dissect a Group A flower, while the other half dissect a Group B flower.

☐2. Use your *own observations* to determine the floral characteristics of your large flower, and record your observations below. Use a dissecting microscope to see details.

Flower 1: Name of your *large* flower: _____

Number of sepals: _____
 Sepals' color and surface texture (smooth or hairs?):

_____
 Are the sepals separate or fused together?

_____

Number of petals: _____
 Petals' color and surface texture:

_____
 Are the petals separate or fused together?

_____

Number of stamens: _____
 Color of the anthers: _____
 Are the stamens separate or fused with *each other*?

_____
 Are the stamens' filaments fused with *the petals*?

_____

Number of pistils: _____
 Number of stigmas per pistil: _____
 Texture of stigmatic surface (smooth, bumpy, hairy):

_____
 Is the ovary superior or inferior? (Refer to Figure 18.2.) _____
 If the ovary is large enough, cut it crosswise:
  How many chambers? _____
  How many ovules per chamber? _____

a. Among the flowers of your species that are being dissected in class, do any deviate from the norm with regard to the number of sepals, petals, stamens, and pistils? Record the norm and the deviation.

b. What flower from the *other* large-flower group was dissected at your table? _____
c. Compare your flower to this flower from the *other group*. Does this other flower have a superior or inferior ovary? _____

☐3. Now pick the largest broccoli bud that you can find on the top of your broccoli stalk; use dissecting needles to dissect the broccoli bud under the dissecting microscope.

Flower 2: Broccoli bud

Number of sepals: _____
 Sepals' color and surface texture:

_____
 Are the sepals separate or fused together?

_____

Note that in a bud, the petals are small and unexpanded, and may look crumpled up.
Number of petals: _____
 Petals' color and surface texture:

_____
 Are the petals separate or fused together?

Note that in a bud, the filaments of the stamens have not yet elongated, and thus are very short. In a broccoli bud, the long anthers and the central pistil are similar in shape, so examine closely.

Number of stamens: _____

  Do all students have this number of stamens in their broccoli bud?

Number of pistils: _____

I.B. UNISEXUAL ("IMPERFECT") FLOWERS

Some plants produce flowers that do not have both male and female structures in the same flower; instead, they have male flowers (having stamens but lacking pistils) and female flowers (having pistils but lacking stamens). Male flowers are also called staminate flowers; female flowers are also called pistillate flowers or carpellate flowers.

In begonia flowers, the sepals and petals are the same color. The female flower of begonias has a bulge with three "wings" beneath the petals (an inferior tricarpellate ovary) and three twisted styles above the petals; the male flower has a cluster of stamens above the petals and no bulge beneath the petals.

MATERIALS

• potted begonias in flower

INVESTIGATION

☐1. Examine the begonia plant and diagnose whether a flower is male or female.
☐2. How many female flowers are on the plant?_____ How many male flowers?_____

NOTE: We will revisit this topic at the end of the lab, after learning more about fruits.

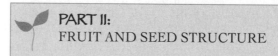

PART II: FRUIT AND SEED STRUCTURE

Fruits Develop From Flowers. A **fruit** is a ripened ovary (or group of ovaries) containing the seeds; in some plants, certain adjacent plant parts are fused with the ripened ovaries. The ripened ovary wall may be dry as in hazelnuts or fleshy as in grapes. In peaches, the inner layer of the ovary wall becomes stony, forming the pit, while the outer portion of the ovary wall becomes fleshy (see Figure 18.3).

The growth and ripening of an ovary wall is largely in response to the **growth hormones** released by the enclosed developing seeds. If you cut open a lopsided apple, you will find no (or almost no) seeds in the underdeveloped side. Exceptions to this general rule are **parthenocarpic** plants (Greek: *parthenos* = virgin; *karpos* = fruit) such as cultivated banana and pineapple; they can produce fruit even in the absence of developing seeds (seedless fruits result).

Although fruits come in all shapes and sizes, they all function in *protection* of the enclosed seeds. Also, the fruits of various plants have evolved structures to aid in *seed dispersal*, such as: dry wings on maple fruits aid in *gliding*, hairs on dandelion aid in *parachuting*, colorful flesh of dogwood aids in *attracting* birds that swallow the fruit (the seed passes through unharmed), and pods of jewelweed explode, flinging the seeds.

Figure 18.3 Peach flower: the inner layer of the ovary wall develops into the stony pit.

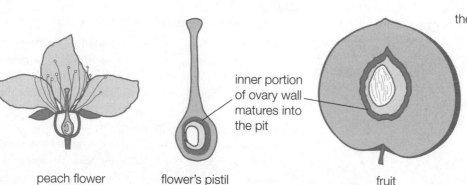

inner portion of ovary wall matures into the pit

peach flower   flower's pistil   fruit

18

MATERIALS

- http://mcintosh.botany.org/bsa/misc/mcintosh/mc-intosh.html (McIntosh Apple Development stages)
- David Attenborough's *The Private Life of Plants*, Vol. 1, "Branching Out" [videocassette]

INVESTIGATIONS

Do the following, at a time indicated by your instructor:

☐ 1. Look at the online interactive version of the McIntosh Apple Development poster on the Botanical Society of America's website (the poster itself can be purchased from them). The poster shows 20 photos of the progression from winter bud to mature apple. Click to view a close-up and descriptive paragraph of each photo.

☐ 2. View *The Private Life of Plants*, Volume 1 video. This video has fantastic footage of all kinds of amazing fruit and seed dispersal methods. As you watch the video, make a list of all the methods of seed dispersal shown.

II.A. WHAT IS INSIDE A SEED?

MATERIALS

- peanuts or soaked bean seeds
- germinating corn, embryo enlarged and barely protruding
- dissecting microscopes and dissecting needles

INVESTIGATIONS

☐ 1. Dissect and find the seed parts described below:

A seed develops from an ovule. The protective seed **coat** (Figure 18.4) around the seed developed from the outer cell layers of the ovule. (*NOTE*: In corn and other grasses, the seed coat and ovary wall fuse into a single structure.)

The **embryo** is a tiny plant consisting of an embryonic *root*, an embryonic *shoot* with tiny leaves, and the cotyledons or "seed leaves" which contain stored food. Peanuts and beans are "dicots" (have two cotyledons), while corn is a "monocot" (has one cotyledon). The embryo is *offspring* tissue, developing from a fertilized egg.

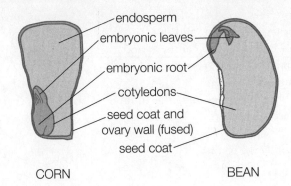

Figure 18.4 Seed structure

The **endosperm** (Greek: *endon* = within; *sperma* = seed) is the tissue outside the embryo but inside the seed coat; like the cotyledons, the endosperm contains stored food for germination. In some species like corn, the endosperm is abundant in the mature seed. In other species like bean or peanut, the endosperm becomes absorbed by the embryo's cotyledons during the course of seed development; thus, at maturity, the bean or peanut seed contains no endosperm but two huge cotyledons that are full of food reserves.

II.B. WHAT ARE YOU EATING?

MATERIALS

- per lab bench, on paper plates: peanuts in the shell, strawberry, raspberry, apple, orange, corn, almond in the shell, and walnut in the shell (in winter, buy a bag of nuts in the shell, and freeze for next time)
- The McIntosh Apple Development Poster, available through the Botanical Society of America
- on front desk, on paper plates: fresh pineapple, coconut, and coconut with husk still intact
- clean paring knives, cleaned bench tops, and paper towels
- nutcrackers and hammer
- dissecting microscopes

INVESTIGATIONS

☐ 1. Work in pairs. For each available specimen, *examine the specimen, read the commentary,* and *consult Figures 18.3 through 18.6.* You will then be able to figure out the fruit parts and answer the questions that follow.

☐ 2. If your instructor permits you to eat the specimens, wait until *after* you have examined the specimens! Also make sure that they have been washed and handled cleanly.

208

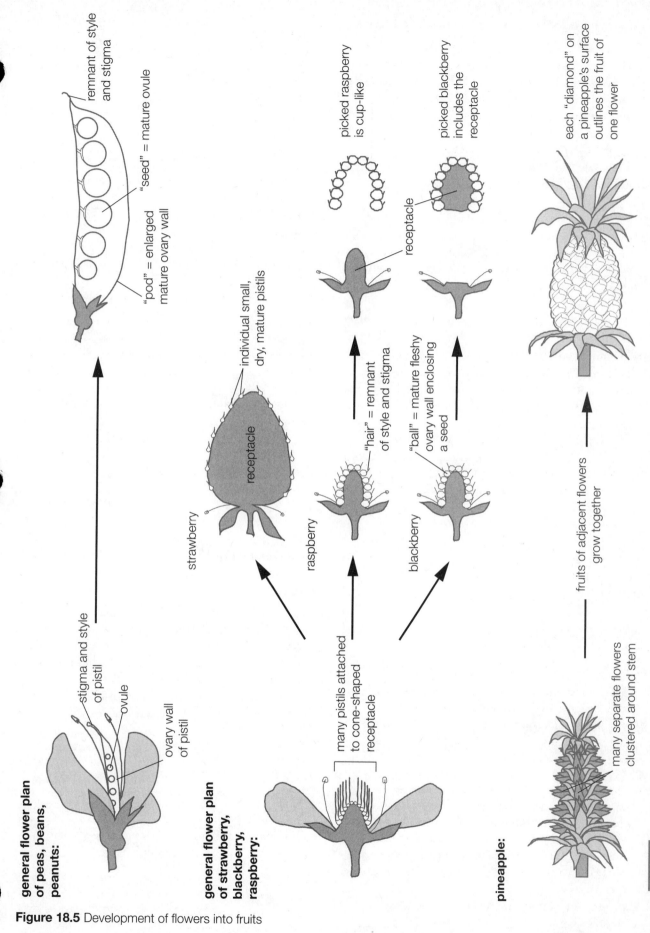

Figure 18.5 Development of flowers into fruits

general flower plan of peas, beans, peanuts:

stigma and style of pistil
ovule
ovary wall of pistil

remnant of style and stigma
"seed" = mature ovule
"pod" = enlarged mature ovary wall

general flower plan of strawberry, blackberry, raspberry:

many pistils attached to cone-shaped receptacle

individual small, dry, mature pistils
receptacle

strawberry

raspberry

"hair" = remnant of style and stigma

picked raspberry is cup-like

receptacle

blackberry

"ball" = mature fleshy ovary wall enclosing a seed

picked blackberry includes the receptacle

pineapple:

many separate flowers clustered around stem

fruits of adjacent flowers grow together

each "diamond" on a pineapple's surface outlines the fruit of one flower

18

Peanut, in the Shell (see Figures 18.4 and 18.5)

Peanuts and peas and beans are in the same plant family; their fruits have the same structure. The mature peanut is found in the ground. The peanut flower is aboveground, and after fertilization, the flower stalk grows downward and pushes the developing peanut fruit under the ground.

The peanut pod is structured and develops just like a bean pod (see Figure 18.5).

From *which flower part* did the rough peanut pod develop: _____

The thin reddish papery covering on the peanut is what botanical structure: _____

What is the little nub connecting the two peanut halves (examine under dissecting scope):_____
- What is the nub's pointy outer end ("Santa's hat")?

_____
- What is the nub's forked V-shaped inner end ("Santa's beard")? _____

What are the two halves of the peanut that you eat?
_____

Why are peanuts also called groundnuts?

How does this fruit's structure promote reproductive success (promote the successful production and dispersal of seeds)?

Strawberry (see Figure 18.5)

From *which flower parts* did the following fruit parts *develop*?
- The collar of green leafy parts: _____
- The red flesh: _____
- The hard little "granules" on the surface:

_____
- The little "hair" attached to each little granule:

_____

Can you find dried-up old stamens associated with your strawberry? _____ How many? _____

How does this fruit's structure promote reproductive success?

Raspberry (see Figure 18.5)

The raspberry flower is similar to the strawberry flower (refer to Figure 18.5). The raspberry consists of a thimble-shaped cluster of fleshy balls that detach from the plug of tissue on the plant when picked.

From *which flower parts* did the following fruit parts *develop*?
- The plug of tissue left behind on the plant when a raspberry is picked: _____
- The little fleshy balls of tissue: _____
- The little hair attached to each fleshy ball:

_____

How does this fruit's structure promote reproductive success?

Apple (see Figure 18.6; interactive poster at http://mcintosh.botany.org/bsa/misc/mcintosh/mcintosh.html)

In the apple flower, *the ovary is surrounded by a mass of tissue* derived from the receptacle and floral tube (the bases of the sepals, petals, and stamens); such an ovary where the sepals, petals, and stamens flare out only from the top of the ovary (see Figures 18.2 and 18.6) is called an *inferior ovary*. The flowers of blueberries and cranberries also have the inferior ovaries.

☐ Slice an apple *crosswise*; observe and determine the following:
- Number of cavities in the core of the ovary (or number of "points to the star"): _____
- Number of seeds per cavity: _____
- If your apple is lopsided, how many seeds per cavity in the small side vs. large side? _____

From *which flower parts* did the flesh that you eat *develop*?
_____

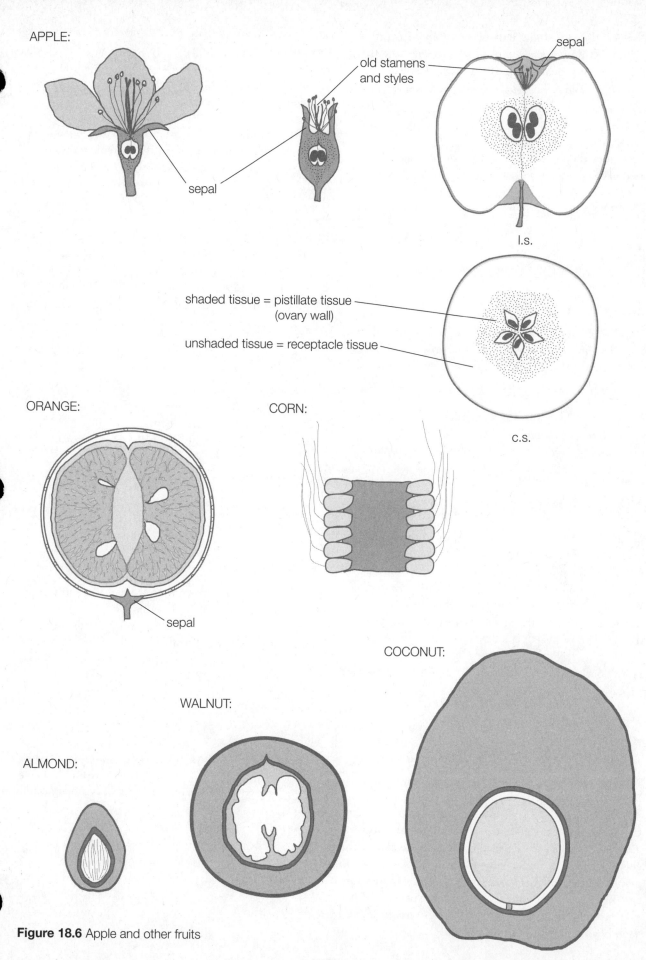

APPLE:

sepal

old stamens
and styles

sepal

l.s.

shaded tissue = pistillate tissue
(ovary wall)

unshaded tissue = receptacle tissue

c.s.

ORANGE:

CORN:

sepal

COCONUT:

WALNUT:

ALMOND:

Figure 18.6 Apple and other fruits

211

☐ Slice the non-stalk half of your apple *lengthwise*. Which old flower parts are present in the little dried clump of stuff located opposite the apple stalk?

- The five triangular tough papery points = _____
- The dried-up strands = _____ and _____

How does the apple's structure promote reproductive success?

Orange (see Figure 18.6)

Did the orange flower have a superior or inferior ovary? (Hint: *Check the location of the sepals* —which are green, small, thick, and fleshy—attached to the fruit stalk.) _____

Each orange segment was a chamber of the ovary and is now filled by *mature, fluid-filled hairs*.
From *which flower parts* did the following fruit parts *develop*?

- The rind: _____
- The juicy flesh: _____

☐ Citrus fruits have many **oil glands** in the rind. Peel part of the rind; fold the peeling over, with the orange surface outermost, and pinch it to sharply kink it and rupture the oil glands; *notice the fine mist of fragrant oils squirting into the air.* You can make the whole room fragrant without buying air freshener! The same is true for lemon peels.

How does this fruit's structure promote reproductive success?

Corn (see Figures 18.4 and 18.6)

Corn has separate male flowers and female flowers. The *male corn flowers* are in the *tassel* at the top of the corn plant. These male flowers have stamens but no pistils. Many *modified leaves* (corn husks) surround an ear of corn, which consists of the *cob* (specialized lateral branch) that is covered with many *female flowers*, each having a pistil but no stamens.

The **corn silk** is made up of the extremely long *styles*, one strand of silk going to one ovary, which is destined to develop into one corn kernel. Thus, each corn kernel came from a different pistil and resulted from a separate fertilization event.

Note that in grasses such as corn, wheat, rice, and so forth, there is one seed in an ovary; fruit maturation in grasses is unusual in that the *ovary wall and the seed coat grow together to form a single structure* surrounding the seed.

Dissect a corn kernel. Can you locate the corn embryo? _____ The endosperm? _____

From which two flower structures did the "coat" on a corn kernel develop: _____ and _____

Which *modified plant structures* are these? Circle the correct choice:

- The cob (to which the kernels are attached) is a modified: branch   leaf   perianth   ovary
- The corn husks are modified:
  leaves   sepals   petals   stamens

Which cob would you expect to have the most kernels: a cob with many or with few silk strands? Why?

The wild varieties of corn do not have the cob tightly encased by many husks. Humans have bred the corn to have these many tight husks. Why would humans consider this feature desirable?

Why do you suppose the wild plants never evolved to having many tight husks? (Related question: Why is corn not a troublesome plant that invades surrounding fields?)

Almond, in the shell (see Figure 18.6)

What does an almond in the shell remind you of? On the almond tree this shell was surrounded by a layer of rather tough flesh—and looked like a tough peach. This almond in the shell is the "pit" or stone of the fruit; crack the shell to see the almond seed inside. Almonds, peaches, plums, and cherries are called the "stone fruits." (See Figure 18.3).

- What is the thin, brown papery covering surrounding the almond? _____
- What is the little nub holding the two halves of the almond together, just like in the peanut? _____
- Since seeds are contained within an ovary wall, then the hard pitted shell that you cracked and the tough flesh that was already removed would all be part of what structure? _____

How does this fruit's structure promote reproductive success?

Walnut, in the shell (see Figure 18.6)

In a walnut, the *inner layer of the ovary wall* becomes very hard and *stony*, while the outer portion of the ovary wall becomes thick and fibrous. (Go back and check your answer for the almond.) The structure of the walnut fruit is similar to the structure of *peaches*, except that the flesh is very tough and fibrous.

The hard "shell" that you crack with a nutcracker developed from which *flower* part: _____

What is the thin, papery covering on the "meat" or "flesh": _____

What are the two convoluted "walnut halves"? (Hint: They are similar to the two halves of a peanut.)

_____

How is the walnut fruit's structure adaptive, promoting reproductive success?

Pineapple (see Figure 18.5)

Note the diamond-shaped sections on the surface of the pineapple. Also note the dry, thin, pointed structure at the base of each of these sections. The pineapple is an example of a *multiple* fruit—the mature ovaries of many flowers being fused into one mass. The stem of the pineapple plant was surrounded by over 100 closely packed flowers, each of which developed into a fruit (one of the diamond-shaped sections). At the base of each flower was a leaflike structure (the now-dry pointy structure on each diamond-shaped section). All these fruits and the central stem fused together to form a pineapple.

Remember that the pineapple is a parthenocarpic fruit—it can develop even without seeds developing within.

When the pineapple is sliced open, look for remnants of the style and stamens in the center of each "diamond." Were you able to find an occasional seed just beneath the remnants of flower parts?_____

From *which plant parts* did the following structures *develop*?

- The central tougher core of the pineapple:

  _____

- The outer fleshy zone (the "ring" in cans):

  _____

Coconut (see Figure 18.6)

The coconut, fruit of the coco palm of tropical beaches, is a very large, almost football-shaped structure; despite its size, it floats. The coconut fruit is also resistant to salt water. The *outer ovary wall* has developed into a thick, tough, football-shaped fibrous husk, surrounding the round, hard, hairy "shell" seen in grocery stores; this round, hard shell is the *inner layer of the thick ovary wall*. You may have seen carved monkeys from the Philippines (see Figure 18.7), made of two coconut fruits with the fibrous husk carved. The hair on the head consists of husk fibers that remain after degradation of the soft husk tissue in which the fibers were embedded.

The coconut **milk** is liquid *endosperm* in which hundreds of nuclei are suspended; as the coconut matures, the endosperm's nuclei settle around the outside and cell membranes and cell walls form around them, producing the white coconut **meat**. Under one of the three "eyes" of the shell is located the single embryo—a little waxy peg in the meat. The papery brown layer on the outside of the meat is the seed coat.

18

From *which flower parts* did the following fruit parts *develop*?

- The fibrous husk: _____
- The round hard shell: _____

What part of the fruit is the coconut meat and milk?

_____

Before the coconut is cracked open, look at the lines, marking out sections; how many carpels (sections) were in the compound pistil of the coconut?_____

☐ Crack the coconut open; beneath *one* of the three "eyes," find the waxy peg embedded in the meat. This peg is the _____ (the beginning of a new palm tree!)

How does this fruit's structure promote reproductive success?

Predicting Pumpkin Production

If you have a pumpkin patch, you *cannot* determine the number of potential pumpkins by counting the *total* number of flowers. Why? Pumpkin plants have separate male flowers and female flowers on the same plant (see Figure 18.8). A pumpkin fruit develops from the pistil of a female flower. Identify a female flower by its bulging inferior ovary at the end of the flower stalk beneath the sepals and petals. Thus, to determine the number of potential pumpkins to pick, you count only the number of *female flowers* present! (This is true also for zucchini, squash, cucumber, and most other members of the gourd family.)

Q3.   You look at your pumpkin patch in the garden and there are seven flowers; upon closer examination you discover that all seven are female flowers. How many pumpkins would you expect to harvest?
(Be careful now . . . remember that "it takes two.")

No Berries on Your Berry Tree?

Some plant species such as sassafras and Osage orange are **dioecious**; that is, an individual plant has either all male flowers or all female flowers. Holly trees are nearly dioecious, producing an occasional **perfect** flower (having both male and female parts). Thus, if you want a holly tree that will produce lots of holly berries for the winter birds and for decoration, you need a tree that is predominantly female. If your holly tree has no or only a few occasional berries, your tree is likely to be of the predominantly male type; it is also possible that you do have a predominantly female tree, but there is no male tree within pollination distance of your tree. (*NOTE*: Plants such as corn that have male flowers and female flowers on the same plant are called **monoecious**.)   (Greek: *monos* = one; *dis* = twice; *oikos* = house)

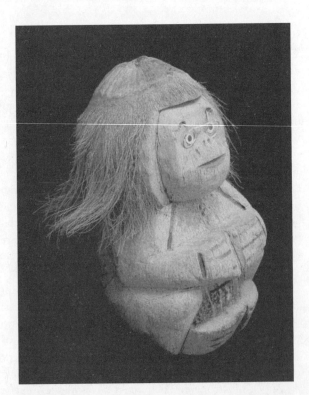

Figure 18.7 Carved monkey made of two coconuts

EXTENDED ASSIGNMENT

Pick a fruit to research! It can be anything from edible fruits, to those pressed for their oil, to those fermented, to those grown for their fiber, dye, medicinal use, and so forth. There is a great deal of fascinating information out there. A good place to start is with one of the various books on "economic botany" (Simpson and Ogorzaly 2001).

Research the structure of the plant that produces that fruit. Find out where and how that plant is grown and pollinated, and how the fruit is harvested, processed, and used. What's the history of the crop? To what geographical area is it native? Which countries are currently the major producers and users? What other plants belong to the same plant family?

Present your findings orally or in writing, as instructed.

LITERATURE CITED

Attenborough, D. 1995. David Attenborough's *The Private Life of Plants*, Vol. 1: "Branching Out" [videocassette]. BBC/Turner Original Productions, Inc., coproducers. Atlanta, GA: Turner Home Entertainment.

Botanical Society of America (BSA). 2001. "McIntosh Apple Development," http://mcintosh.botany.org/bsa/misc/mcintosh/mcintosh.html (accessed February 19, 2006). St. Louis, MO: Education Committee, Botanical Society of America.

Simpson, B. B., and M. C. Ogorzaly. 2001. *Economic Botany: Plants in Our World*, 3rd ed. Boston: McGraw-Hill.

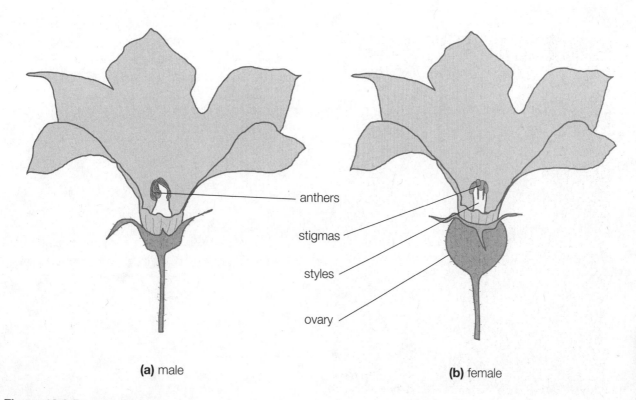

(a) male   (b) female

Figure 18.8 Pumpkin flowers

18

PLANT INTERACTIONS

LAB

I. Microhabitats
II. Allelopathy: "chemical warfare" among plants
III. Invasive aliens! (information retrieval)

John Donne wrote in 1624 that "no man is an island"; this saying also holds true for plants. A plant is variously affected by the physical environment as well as by neighboring plants and other living organisms. Part I of this lab investigates temperature in microhabitats. Part II investigates claims of allelopathic interactions between plants. Part III investigates alien plant species that are designated as invasive in your area. Your instructor may have the class do only certain parts of this lab or may have different student groups do different parts and then report their results to the class.

PART I:
MICROHABITATS

OBJECTIVES

1. Determine the vertical or horizontal temperature gradient at a particular microhabitat.
2. Practice the scientific method by properly designing and performing your own experiment.

BACKGROUND

Have you ever been outside on a chilly late-winter day and observed seedlings already growing and insects crawling around on the ground, despite the cold? Have you ever noticed plants growing between boulders on a cold mountaintop? Have you seen young cacti growing under desert trees rather than out in the open? Have you noticed different vegetation on the north slope as compared to the south slope of a canyon or valley? You are probably familiar with the difference in temperature and light conditions of a north-facing windowsill versus those of a south-facing windowsill.

There are various locations, called **microhabitats**, that can have different pH and nutrient conditions and/or microclimates (temperature, humidity, light, and air circulation) than those in the adjacent area. Temperature taken 1.5 meters above the ground gives an overall indication about a particular environment, but it is not necessarily representative of all locations within that environment. Certain organisms may be able to exist in the conditions of a microhabitat, but not of the adjacent area. Even the sunny and the shady side of a single plant are microclimates that produce leaves differing in structure (see Lab 11 on "Leaves").

In this lab, you will investigate the temperatures in various locations within an environment in order to become aware of the heterogeneity in an environment, which in turn can cause heterogeneous plant distribution.

MATERIALS

- thermometers, non-mercury; glass thermometers should each have an open case or armor to protect them from breakage in the field (various thermometers with open cases are available from Carolina, Ward's, and other suppliers).
- meter sticks or tape measure
- whistle, cell phones, or two-way radios (walkie-talkies)

INVESTIGATIONS

☐ 1. Working in groups, select a question to investigate, such as one of the following:
- In a forest or grassland on a chilly day, how does the temperature change vertically, from 3 cm beneath the soil to 1 meter above the soil? (Be sure to include readings at the soil surface, in the litter layer, and at several other positions above.)

19

- If you have both a forest and grassland available to you, how do they compare in their vertical temperature profile?
- How does the temperature of the soil or at the soil surface change as you progress from the stem of a shrub into the surrounding desert or grassland or dunes?
- How does the temperature change as you progress from a rock crevice out into the open?
- How does the temperature profile change from early afternoon to late afternoon?
- How does the temperature profile taken on a chilly day compare to that taken on a hot day?
- How does a vertical temperature profile of mulched and unmulched flower beds differ?
- How much does a passing cloud affect the temperature profile at the microhabitats?

☐2. Locate appropriate microhabitat sites for your investigation.

☐3. Record your question and hypothesis on Data Sheet 1.

☐4. Consider the following when designing your experiment:

- Remember that the hallmarks of good experimental design are *replication*, a *control*, and a *single variable*.
- Will you take the temperature at the various vertically or horizontally arranged points of your site simultaneously or sequentially? How stable is the temperature likely to be throughout the course of your lab period?
- You may decide to have several groups gathering data for the same experiment, in order to gather more data and/or to gather it simultaneously.
- How long will you allow your thermometer to equilibrate (reach a stable reading) at each position?
- Should the thermometer be shaded while it is taking a reading? Why or why not?
- If temperatures are to be recorded at different sites at exactly the same moment, how will you signal to record the readings—by whistle, cell phone, or two-way radio?
- When measuring distances vertically, what number will you assign to the soil surface, and what numbers to underground positions as compared to aboveground positions—so that you can graph your data later?

☐5. Will you be using more than one thermometer in your experiment? Do you plan to pool your data with those of another group or compare your data to those of another group? Then be sure to check your thermometers against each other to make sure that they are recording exactly the same temperature in a particular location. (Is one of the thermometers consistently recording higher or lower?)

☐6. Present your experimental design to the class for discussion and suggestions for improvement.

☐7. After the discussion, revise your experimental design.

☐8. Conduct the experiment. *NOTE: As always, when outdoors, be alert for such things as venomous animals and poison ivy.* Record your site descriptions and temperature readings at the various times and positions within each site, on Data Sheet 1.

☐9. Represent your findings graphically, plotting temperature against distance (consult Appendix 2).

☐10. Present your results, orally or in writing, and discuss their implications for plant growth and survival in the microhabitat through the seasons. Discuss other factors besides temperature that could be different in a microhabitat than in the surrounding area, influencing the organisms that occupy the microhabitat.

DATA SHEET 1. Temperature in Microhabitats

Your question:

Your hypothesis:

Experimental design:

Temperatures recorded at different positions within each microhabitat site, and time at which these temperatures were recorded:

Position (distance in cm)	Site 1 time | temp	Site 2 time | temp	Site 3 time | temp	Site 4 time | temp	Site 5 time | temp

Description of sites:

Site 1:

Site 2:

Site 3:

Site 4:

Site 5:

19

PART II:
ALLELOPATHY: "CHEMICAL WARFARE" AMONG PLANTS

OBJECTIVES

1. Test a claim, found in a tertiary literature source, of allelopathic interactions between two species.
2. Practice the scientific method by properly designing and performing your own experiment.
3. Conduct a literature search for primary sources.

BACKGROUND

Plants are variously affected by neighboring plants. A tall plant casts shade on shorter plants. A vine growing on plants shades them and sometimes even kills them due to the shade and/or twining stems. In dense growth, humidity is higher due to the reduced air circulation and evaporation. Certain plants release various chemicals into the soil and air that can have beneficial effects on certain other plants; this is the basis of **companion planting**, whereby gardeners plant certain kinds of plants together because their association appears to improve their growth. The opposite situation also occurs: Some plants release chemicals that are detrimental to the growth of other plants nearby; this phenomenon is referred to as **allelopathy**.

Allelopathy is a very active area of research. Topics under investigation include such things as competition in natural ecosystems, weed and crop interactions, effects of exotic species on native vegetation, isolation and identification of the allelopathic chemicals, and so on. In the tertiary literature (written for the nonscientist, with no literature citations given), you will sometimes come across various claims of allelopathic interactions and wonder what scientific data are backing it up. For example:

- "Black walnut roots cause tomato plants to wilt and die" (McClure 1994).
- "Of all the plants with allelopathic properties, black walnut (*Juglans nigra*) is probably the most notorious. . . . Rain dissolves juglone from the leaves and washes it down into the rooting area, killing or stunting many kinds of plants" (McClure 1994).
- "Rye seedlings are strongly allelopathic but grow more benign as they mature" (McClure 1994).
- "Garden mums (*Chrysanthemum* × *morifolium*) . . . produce an allelopathic compound in their leaves.

When the compound washes out of the leaves . . . it prevents lettuce seeds from germinating" (McClure 1994).
- "Broccoli and cabbage are widely allelopathic" (McClure 1994).
- "Coriander when grown near anise helps anise seeds to germinate and to grow . . . coriander has the opposite effect on fennel and will hinder germination of the seeds" (Hemphill and Hemphill 1983).
- "Savory seeds inhibit the germination of other seeds when planted nearby" (Hemphill and Hemphill 1983).

You will choose one of these claims or some other claim of allelopathic interaction to investigate. If you are in the desert, you could choose one of the regularly spaced shrubs such as creosote bush to test for allelopathic effects.

MATERIALS

- seeds such as: rye, coriander, anise, fennel, savory, lettuce, radish, and tomato
- plant material:
  - black walnut (shells, leaves, or roots)
  - leaves of garden mums (*Chrysanthemum* × *morifolium*)
  - broccoli and cabbage—vegetative parts
- distilled water
- mortars and pestles, quartz sand
- paper towels
- petri dishes—several sleeves
- Parafilm M® and scissors
- containers: plastic pots, recycled yogurt cups, the bottom half of plastic soft drink bottles, or other containers
- large nails and hammers for making drainage holes in the bottom of recycled plastic containers
- potting soil in tub; if dry, add water and stir with trowel to moisten before use
- trowels, labels, and felt-tip markers

INVESTIGATIONS

☐1. Working in groups, select the allelopathic interaction that you wish to investigate. Record your question and hypothesis on Data Sheet 2.
☐2. Check germination times on the seed packets. Note that this time is usually specified for seeds growing outdoors, under fluctuating temperature and humidity conditions. Seeds grown under constant warmth and humidity generally germinate quite a bit faster.

☐3. *To make a plant extract:* Either rinse the plant material with water or grind the plant material with quartz sand and water in a mortar and pestle. Use the resulting solution to water your seeds and/or seedlings.

☐4. *To be able to observe your seed germination directly:* Fold a paper towel into a square and cut off the corners so that it fits into a petri dish. This will serve as a wet platform that sits in a pool of solution in the dish and wicks up the solution to keep the seeds on top uniformly moist throughout the experiment. *To prevent evaporation,* seal the edges of each closed dish with a 1.5 cm wide strip of Parafilm. (To do this, hold the beginning of the Parafilm strip against the edge of the dish and pull the strip with the other hand to stretch it while wrapping it 1 1/4 times around the edge of the dish; then anchor the stretched Parafilm end by pressing it onto the stretched Parafilm on the dish—stretched Parafilm is tacky and will stick to stretched Parafilm).

☐5. *To test seed germination in soil:* Make drainage holes in the bottom of your plastic containers. Add soil to the containers and gently tamp it down; repeat until the soil level is 1–2 cm below the rim. After planting seeds, carefully keep the seeds moist with the appropriate solution.

☐6. Should you use sterile materials and surface-sterilize the seeds to eliminate mold? Would it matter? You could find out by having one student group use surface-sterilized seeds and another group use regular seeds. (Seeds can be surface-sterilized by putting them in 10% bleach solution for 5 minutes and then rinsing them thoroughly with water for 5 minutes. *NOTE: Use safety glasses. Bleach is corrosive; do not get in eyes, on skin, or on clothing.*)

☐7. Consider the following when designing your experiment:
- Remember that the hallmarks of good experimental design are *replication*, a *control* (untreated seeds), and a *single variable*.
- You may decide to have several groups testing the same hypothesis to generate more data, which can then be pooled for analysis.
- How will you make sure that the plant extract or rinse that you apply to all of your seeds and/or groups of seedlings at a particular time is the same concentration and quantity?
- What numerical data will you take, and when?
- To be able to determine whether there is a *statistically* significant difference between germination of treated and untreated seeds, for example, you would want numerous petri dishes for the treated group and also for the untreated group, with each dish containing the same number of seeds (10 perhaps); then you could perform the **two-sample t-test** (see Appendix 3) comparing the number of germinated seeds per dish in the untreated group of dishes to the number of germinated seeds per dish in the treated group of dishes.

☐8. Present your experimental design to the class for discussion and suggestions for improvement.

☐9. After the discussion, revise your experimental design and conduct the experiment. Record your data on Data Sheet 2.

☐10. Search the scientific literature for original research papers (primary literature) on your plant; use as keywords **allelopathy** and the genus of the scientific name of your plant. Check the online database AGRICOLA at http://agricola.nal.usda.gov/ (anyone can access this database; your tax dollars at work!) as well as BIOSIS if your institution has access to it. AGRICOLA covers various agricultural journals that BIOSIS does not cover.

☐11. Summarize your results in tables and graphs (refer to Appendix 2) and analyze your date (refer to Appendix 3).

☐12. Prepare an oral or written report, as instructed, presenting and analyzing your data and comparing it to the literature. Was there a significant difference between your control and experimental groups? Based on your findings, can you definitely state that the material under investigation does or does not have allelopathic properties? What would be required for you to give a definitive answer?

EXTENDED ASSIGNMENT

Based on your results, design another experiment to improve on your data.

DATA SHEET 2. Allelopathy Experiment

Your question:

Your hypothesis:

Experimental design:

PART III:
INVASIVE ALIENS!
(information retrieval)

OBJECTIVES

1. Determine which alien plant species are invasive in your geographic area.
2. Research the history and current status of the invasive alien plants in your geographic area.

BACKGROUND

Plant species that are not native to an ecosystem are called **alien** or **exotic** species. Many of your garden plants are aliens. An alien species, in the absence of its usual controls (diseases and insects present in its native area), may become **invasive**—spread rapidly and outcompete the native vegetation. Loss of native vegetation or invasion of crop fields is a major problem in some areas; in that case, the alien is often called a **noxious weed**. If native vegetation is replaced with an invasive alien of lesser food value and of different growth habit, the wildlife of the area can be significantly affected.

Just because an alien plant is legally sold in your area does *not* mean that it will be noninvasive. This plant may already have become invasive in another part of the country, but not yet in your area. Therefore, look at how the plant "behaves" in other areas of the country. By the time a law is passed to make the sale of an alien plant species illegal in your area, there is generally already a problem in your area.

Each country is struggling with invasive alien plants (as well as fungi, animals, and other organisms) that are native elsewhere and that were originally introduced either accidentally or for a particular purpose. There is a reason behind the laws that restrict and regulate the importation of plants and other living materials.

INVESTIGATIONS

☐1. Determine the most important invasive alien plant species in your area by searching the Internet and the print literature as well as consulting governmental agencies. Search the Internet with the search engine of your choice, using keywords such as **alien plants** and the name of your country; select several of the websites and work your way through them

until you get to information for your geographic region.

☐2. Compare the information presented by several websites (governmental and other).

☐3. Select the invasive aliens of greatest concern in your area and research them. Your instructor may have each student group research a different invasive alien species.

☐4. Determine the native range of the species, its range as an alien, the reason for its introduction to your area, its ecological effects (negative as well as positive), the geographic locations where it is considered particularly troublesome, the control measures being attempted, and the effectiveness of and problems associated with the control measures.

☐5. Summarize your findings in an oral presentation or in writing, as indicated by your instructor.

☐6. Your instructor may take you on a field trip to view areas being invaded by alien plant species. Don't be surprised if the field trip is on campus.

EXTENDED INVESTIGATION:
Reproductive Potential

What is the reproductive potential of a particular species of invasive plant?

☐1. Working in groups, select an invasive alien (often called a common weed) in your area.

☐2. Determine what a "typical" individual of this species would be. Find several such individuals.

☐3. How many seeds does a typical individual produce?

Note that invasive aliens often produce many seeds; therefore, you will first have to discuss potential sampling methods for estimating the total number of seeds produced by an individual. Determine the total number of seeds produced by each of your typical individuals and take the mean (average) for these typical individuals.

☐4. If all seeds of a typical individual were dispersed, grew, and reproduced generation after generation, how many individuals will there be in the population each year for the next 10 years?
Graph this information.

☐5. How long would it take for that species to cover a hectare? A square kilometer?
To do this determine the area that one individual covers. Determine whether this species is an annual, biennial (flowers in its second year), or perennial (how old is it when it first reproduces?). Then do the calculations.

☐6. What would these calculations look like if only 10% (or 1%) of the seeds that are produced, actually survive to grow into a plant?

LITERATURE CITED

Hemphill, J., and R. Hemphill. 1984. *Herbs: Their Cultivation and Usage*. Poole, Dorset, England: Blandford Press.

McClure, S. 1994. *Companion Planting*. Emmaus, PA: Rodale Press.

USDA, National Agricultural Library. 2005. NAL Catalog, http://agricola.nal.usda.gov/ (accessed February 19, 2006).

APPENDIX

THE METRIC SYSTEM

I. Exponents and scientific notation
II. The metric system
III. Converting units
IV. Centigrade (Celsius) and Fahrenheit interconversions
V. Test yourself!

For some interesting reading, see the Dictionary of Units of Measurement website:

Rowlett, R. 2006. "Dictionary of Units of Measurement," http://www.unc.edu/~rowlett/units/index.html (last accessed March 6, 2006). Chapel Hill, NC: Center for Mathematics and Science Education, University of North Carolina.

I. EXPONENTS AND SCIENTIFIC NOTATION

LARGE number

10^3	=	$10 \times 10 \times 10$	=	1000.
10^2	=	10×10	=	100.
10^1	=	10	=	10.
10^0	=		=	1.
10^{-1}	=	$1/(10^1)$	=	0.1
10^{-2}	=	$1/(10^2)$	=	0.01
10^{-3}	=	$1/(10^3)$	=	0.001

SMALL number

Arithmetic with Exponents

$10^a \times 10^b = 10^{(a+b)}$ Example: $10^3 \times 10^4 = 10^7$
                                                    $(1000 \times 10000 = 10,000,000.)$

$\dfrac{10^a}{10^b} = 10^a \times 10^{-b} = 10^{(a-b)}$ Example: $\dfrac{10^4}{10^3} = 10^4 \times 10^{-3} = 10^{(4-3)} = 10.$

A number in **scientific notation** has *only one* **nonzero digit** *to the left of the decimal.*

Examples of Conversion *from* Scientific Notation

$3.2 \times 10^5 = 3.2 \times 100,000 = 320,000.$ (moved decimal point 5 places to the right)

$4 \times 10^{-2} = 4 \times (1/100) = 4/100 = 0.04$ (moved decimal point 2 places to the left)
Note that for clarity, a zero is written to the left of the decimal of a number that is less than one.

Examples of Conversion *to* Scientific Notation

$320,000. = 3.2 \times 100,000. = 3.2 \times 10^5$

$0.00072 = 7.2/10000 = 7.2/(10^4) = 7.2 \times 10^{-4}$

$567 \times 10^5 = (5.67 \times 10^2) \times 10^5 = 5.67 \times 10^7$

$5243.2 \times 10^{-6} = (5.2342 \times 10^3) \times 10^{-6} = 5.2342 \times 10^{(3-6)} = 5.2432 \times 10^{-3}$

II. THE METRIC SYSTEM

Moving within the metric system involves simply moving the decimal point, whereas moving within the U.S. system involves laborious calculations. For example:

$$3,200,000. \text{ millimeters} = 3,200. \text{ meters} = 3.2 \text{ kilometers}$$

$$3,200,000. \text{ inches} = 266,666.7 \text{ feet} = 88,888.9 \text{ yards} = 50.5 \text{ miles}$$

Some relationships between the U.S. and metric systems:

1 inch = 2.54 centimeters
1 yard ≈ 0.91 meter
1 statute mile ≈ 1.61 kilometer
1 square mile ≈ 259 hectare = 2.59 km² (square kilometer)

1 quart (U.S. liquid) ≈ 0.95 liter
1 ounce (avoirdupois) ≈ 28.35 grams
1 pound (avoirdupois) ≈ 0.45 kilograms

Prefixes in the Metric System Have a Specific Meaning

Prefix	Power	Unit	Symbol	Grams	Value
Tera-	10^{12}	1 Teragram	= 1 Tg	= 10^{12} grams	= 1,000,000,000,000. grams
Giga-	10^{9}	1 Gigagram	= 1 Gg	= 10^{9} grams	= 1,000,000,000. grams
Mega-	10^{6}	1 Megagram	= 1 Mg	= 10^{6} grams	= 1,000,000. grams
kilo-	10^{3}	1 kilogram	= 1 kg	= 10^{3} grams	= 1,000. grams
hecto-	10^{2}	1 hectogram	= 1 hg	= 10^{2} grams	= 100. grams
deka-	10^{1}	1 dekagram	= 1 dkg	= 10^{1} grams	= 10. grams

the basic unit: meter (m), liter (L), or gram (g)

Prefix	Power	Unit	Symbol	Meter	Value
deci-	10^{-1}	1 decimeter	= 1 dm	= 10^{-1} meter	= 0.1 meter
centi-	10^{-2}	1 centimeter	= 1 cm	= 10^{-2} meter	= 0.01 meter
milli-	10^{-3}	1 millimeter	= 1 mm	= 10^{-3} meter	= 0.001 meter
micro-	10^{-6}	1 micrometer	= 1 μm	= 10^{-6} meter	= 0.000001 meter
nano-	10^{-9}	1 nanometer	= 1 nm	= 10^{-9} meter	= 0.000000001 meter
pico-	10^{-12}	1 picometer	= 1 pm	= 10^{-12} meter	= 0.000000000001 meter

LENGTH

kilometer (km)
   1 *kilo*meter = 1000 meters = 10^3 m

meter (m)

centimeter (cm)
   1 *centi*meter = one hundredth of a meter = 10^{-2} m
   thus 100 centimeters = 1 meter

millimeter (mm)
   1 *milli*meter = one thousandth of a meter = 10^{-3} m
   thus 1000 mm = 1 m
   10 mm = 1 cm

micrometer (μm)
   1 *micro*meter = one millionth of a meter = 10^{-6} m
   1000 μm = 1 mm
   1000 mm = 1 m

AREA

square meter (m²)

$$1 \text{ m}^2 = 1 \text{ meter} \times 1 \text{ meter} = 100 \text{ cm} \times 100 \text{ cm} = 10,000. \text{ cm}^2$$

are (a) (This unit is rarely used.)

$$1 \text{ are} = 10 \text{ m} \times 10 \text{ m} = 100 \text{ m}^2$$

hectare (ha)

Hecto- means 100.

$$1 \text{ } hect\text{are} = 100 \text{ ares} = 100 \times (100 \text{ m}^2) = 10,000. \text{ m}^2 = 100 \text{ meter} \times 100 \text{ meter}$$
(1 ha ≈ 2.47 acres)

square kilometer (km²)

Kilo- means 1000. A *kilo*meter is 1000 meters.

$$1 \text{ square kilometer is } 1000 \text{ meter} \times 1000 \text{ meter} = 1,000,000. \text{ m}^2$$
(1 km² = 100 hectares)
(1 km² ≈ 0.386 square mile)

WEIGHT

kilogram (kg)

$$1 \text{ } kilo\text{gram} = 1000 \text{ grams} = 10^3 \text{ g}$$

gram (g)

milligram (mg)

$$1 \text{ } milli\text{gram} = \text{one thousandth of a gram} = 10^{-3} \text{ g}$$
thus 1000 mg = 1 g

microgram (μg)

$$1 \text{ } micro\text{gram} = \text{one millionth of a gram} = 10^{-6} \text{ g}$$
1000 μg = 1 mg
1000 mg = 1 g

VOLUME

Liter (L)

milliliter (ml)

$$1 \text{ } milli\text{liter} = \text{one thousandth of a Liter} = 10^{-3} \text{ L}$$
thus 1000 ml = 1 L

microliter (μl)

$$1 \text{ } micro\text{liter} = \text{one millionth of a liter} = 10^{-6} \text{ L}$$
1000 μl = 1 ml
1000 ml = 1 L

III. CONVERTING UNITS

Remember:
- You can substitute anything with its equivalent term.
- You can multiply anything with a fraction that is equal to the value of 1.
- Cancel out the units. (A unit that appears in the numerator and the denominator can cancel.)

The strategy:
- Multiply your value with fractions equivalent to 1.
- Write the fractions in such a way that you can cancel out units.
- Continue doing this until you have the desired units left over.

Examples:

1200 ml = ? Liters

Remember that 1 Liter = 1000 ml.

Thus the fractions (1 L/1000 ml) and (1000 ml/1 L) are each equivalent to the value of 1.

Multiply 1200 ml with the fraction (1 L/1000 ml) because that will allow you to cancel out the ml units, leaving Liters.

$$1200 \text{ ml} \left(\frac{1 \text{ L}}{1000 \text{ ml}}\right) = \frac{1200 \text{ L}}{1000} = 1.2 \text{ Liters}$$

3,500,000 cm = ? km

Remember that 100 cm = 1 meter, and that 1000 meters = 1 km.

$$3,500,000 \text{ cm} \left(\frac{1 \text{ m}}{100 \text{ cm}}\right)\left(\frac{1 \text{ km}}{1000 \text{ m}}\right) = \frac{3,500,000 \text{ km}}{100,000} = 35. \text{ km}$$

420,000,000 cm² = ? hectares

Remember that 1 m² = 100 cm × 100 cm = 10,000 cm².

Remember that 1 hectare = 100 m × 100 m = 10,000 m².

$$420,000,000 \text{ cm}^2 = \left(\frac{1 \text{ m}^2}{10,000 \text{ cm}^2}\right)\left(\frac{1 \text{ hectare}}{10,000 \text{ m}^2}\right) = \frac{420,000,000 \text{ ha}}{100,000,000} = 4.2 \text{ ha}$$

0.000043 kg = ? mg

Remember that 1 kg = 1000 g, and that 1 g = 1000 mg.

$$0.000043 \text{ kg} \left(\frac{1000 \text{ g}}{1 \text{ kg}}\right)\left(\frac{1000 \text{ mg}}{1 \text{ g}}\right) = 0.000043(10^6) \text{ mg} = 43. \text{ mg}$$

IV. CENTIGRADE (CELSIUS) AND FAHRENHEIT INTERCONVERSIONS

Centi- means hundredth (10^{-2} or 1/100). The centigrade scale has 100 units between the freezing and boiling points of water. The Centigrade scale is also called the Celsius scale, after Anders Celsius (1701–1744), the Swedish astronomer who first described it in 1742. The Fahrenheit scale is named after the German physicist Daniel Fahrenheit (1686–1736), who invented the mercury thermometer and devised this scale. The Fahrenheit scale has 180 degrees between the freezing point (32°F) and the boiling point (212°F) of water.

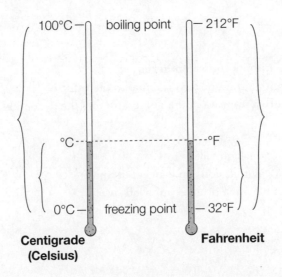

For a particular temperature, the following proportion has the same value for each thermometer regardless of what scale, Centigrade or Fahrenheit, is printed alongside it.

$$\frac{\textbf{distance (units) from the mercury front to freezing point}}{\textbf{distance (units) from boiling point to freezing point}} = \frac{°C - 0°}{100° - 0°} = \frac{°F - 32°}{212° - 32°}$$

Remembering this equality, you can derive the interconversion formulas whenever you need them.

$$\frac{°C - 0}{100 - 0} = \frac{°F - 32}{212 - 32}$$

$$\frac{°C}{100} = \frac{°F - 32}{180}$$

$$°C = \frac{100}{180}\left(°F - 32\right) \qquad -OR- \qquad \frac{180}{100}\,(°C) = °F - 32$$

$$°C = \frac{5}{9}\left(°F - 32\right) \qquad\qquad\qquad \left[\frac{9}{5}\,(°C)\right] + 32 = °F$$

Examples:

Body temperature 98.6°F = ? °C

$$°C = \frac{5}{9}\left(98.6°F - 32\right) = \frac{5}{9}\,(66.6) = 37°C$$

Room temperature 70°F = ? °C

$$°C = \frac{5}{9}\left(70°F - 32\right) = \frac{5}{9}\,(38) = 21.11°C$$

28°C = ? °F

$$°F = \left[\frac{9}{5}\,(28°C)\right] + 32 = 50.4 + 32 = 82.4°F \approx 82°F$$

(Note the peculiarity that for *this* temperature, if you simply *reverse* the number, you get the approximate value on the other scale! Another temperature for which this occurs is $16°C \approx 61°F$.)

It comes in very handy to know the above equivalences as quick reference points.

–40°C = ? °F

$$°F = \left[\frac{9}{5}\,(-40°C)\right] + 32 = -40°F \qquad \text{Same number exactly on both scales! Very cold!}$$

V. TEST YOURSELF!

Put into decimal notation (no exponents):

1. 4.732×10^8

2. 56.4×10^4

3. $321. \times 10^{-5}$

4. 10^{-3}

Put into scientific notation:

5. 78,960,000. =

6. 4900. =

7. 0.03 =

8. 0.000046 =

Moving *within* the metric system: Convert to more suitable metric units that would result in the fewest nonzero numbers to the *left* of the decimal point. Do not convert to scientific notation.

9. 5670. milliliters

10. 5420. grams

11. 0.000005 Liters

12. 23400. meters

13. 125000. kilobytes

14. 4,300,000. watts

15. 78230. micrograms

16. 0.0432 megabytes

17. 0.000023 kg

Calculate area:

18. How many m² is an area 90 cm × 50 cm?

19. How many hectares is a field 100 m × 1000 m?

20. How many hectares is a 1 km × 10 km tract of land?

Temperature conversion: Fahrenheit ⇔ Centigrade (without a calculator's conversion key!):

21. 90°F =

22. 60°F =

23. 20°F =

24. 0°F =

25. 40°C =

26. 10°C =

27. –10°C =

28. –80°C (lab deep freezers) =

1. 473,200,000
2. 564,000
3. 0.00321
4. 0.001
5. 7.896×10^7
6. 4.9×10^3
7. 3×10^{-2}
8. 4.6×10^{-5}
9. 5.67 Liters
10. 5.42 kilograms
11. 5 microliters
12. 23.4 kilometers
13. 125 megabytes
14. 4.3 megawatts
15. 78.23 milligrams
16. 43.2 kilobytes
17. 23 milligrams
18. 0.45 square meters
19. 10 hectares
20. 1000 hectares
21. 32.2°C
22. 15.6°C
23. −6.7°C
24. −17.8°C
25. 104°F
26. 50°F
27. 14°F
28. −112°F

Appendix 1

DATA PRESENTATION

I. Experiments and factors investigated
II. Tables and graphs
III. Rate determination

I. EXPERIMENTS AND FACTORS INVESTIGATED

Experiments often take the following form when investigating a single factor: Effect of (*factor investigated*) on (*characteristic measured*) in (*organism or cell type*). Examples:

- Effect of *seed age* on *germination rate* in *radish seeds*
- Effect of *various growing seasons* on *shoot growth* in *white oak*

Some experiments study the effect of the presence vs. absence of a factor:

- Effect of *fertilizer* on the *seed weight* of *lima beans*

(In such an experiment, the **control** is the group of plants growing in the **absence** of the factor.)

Some experiments study two different factors:

- Effect of various light treatments (factor 1) on percent seed germination in three varieties of lettuce (factor 2)

II. TABLES AND GRAPHS

- Tables are labeled "Tables," and they are numbered sequentially in the report.
- Graphs, charts, drawings, and so on are labeled "Figures" and numbered sequentially in the report.
- Each table and figure has a descriptive title. The table or figure with its title should be *understandable on its own* without the reader having to consult the text.

TABLES

Tables allow the reader to compare exact numerical values.

Table 1. Average yearly twig growth (N = 20) on a sugar maple in 2003–5

year	avg. length (cm)
2003	4.3
2004	5.9
2005	2.1

Table 2. Average yearly twig growth on two tree species in 2003–5

year	average length (cm), N = 20	
	silver maple	white oak
2003	12.5	9.2
2004	16.9	11.4
2005	9.6	5.6

A good table should:

- Have a table number, followed by a descriptive title that specifies:
  - characteristic measured.
  - organism or material involved.
  - factor(s) investigated.
- Indicate the number (N) of data points from which a percentage or average was calculated.
- Have a heading for each column and row.
- Have the measurements or calculated values down the columns. (Numbers are easiest to compare when they are vertically arranged.)
- Indicate the units for the measurements or calculated values (use the metric system).
- Have horizontal lines positioned as shown in Tables 1 and 2 above; tables have no vertical lines.

GRAPHS

Graphs provide an overview; they allow the reader to quickly see trends and make comparisons. A graph has two axes and presents a *measured quantity* versus *the condition over which it is measured.*
Examples:
- Number of seeds germinated versus time
- Final % seed germination versus salt concentrations

The **independent variable** is the range of conditions over which the measurements are taken; this is plotted on the horizontal axis (x-axis). The **dependent variable** is the measured quantity; this is plotted on the vertical axis (y-axis).

Common types of graphs:

1. **Line graph** (called "scatter graph" in Microsoft® Excel® Chart Wizard)
   - Use a line graph when the values of the independent variable are *numerically related.*

   Examples:
   - final percent germination at various *salt concentrations* (draw one curve: % germination versus salt concentrations)
   - percent seed germination over *time* at various temperatures (for each temperature, draw a curve: % germination versus time)

2. **Bar graph** (called "column graph" in Microsoft® Excel® Chart Wizard)
   - Use a bar graph when the values of the independent variable are not numerically related.

   Examples:
   - final percent seed germination for *several varieties of corn*
   - final percent seed germination of *several plant species* in light versus dark

3. **Histogram** (located in the Microsoft® Excel® "Tools" menu, under "Data Analysis"; if Data Analysis does not appear, then click on "Add-Ins" and install "Analysis ToolPak.")
   - Use this type of bar graph to show the distribution of data in a sample.
   - To compare two samples, put the two histograms on the same axes.

   Examples:
   - Seed weights from lima bean plants grown with and without fertilizer
   - Twig growth of a sugar maple (*Acer saccharum*) in 2001 and 2002

Line Graph:

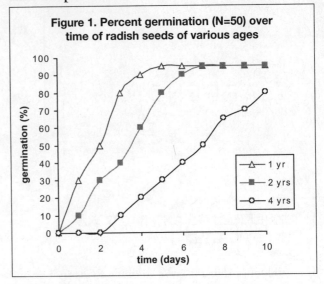

Figure 1. Percent germination (N=50) over time of radish seeds of various ages

Bar Graph:

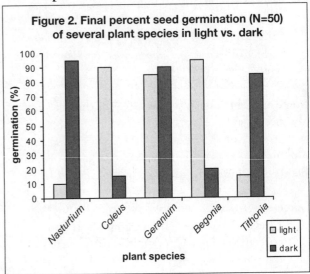

Figure 2. Final percent seed germination (N=50) of several plant species in light vs. dark

Histogram:

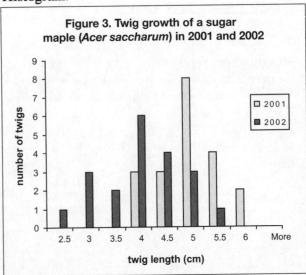

Figure 3. Twig growth of a sugar maple (*Acer saccharum*) in 2001 and 2002

NOTE: A bar includes data up to and *including* that bin value.

A good graph should:

- Have the figure number, followed by a descriptive title that specifies:
  - characteristic measured.
  - organism or material involved.
  - factor(s) investigated.
- Indicate the number (N) of data points from which a percentage or average was calculated
- Have the dependent variable (the measured quantity) on the vertical axis and the independent variable over which range the measurement is taken on the horizontal axis.
- Have an uncluttered, clearly marked and labeled scale on each axis.
- Have each axis labeled, including the metric units in parentheses. Example: Twig length (cm).
- Have clearly marked data points, with symbols for one curve being clearly different from the symbols used for another curve on the same graph.
- In the case of a report with several graphs having the *same* legend entries, use the same symbol and color for a particular entry in *each* graph. Example: Use a solid black circle for the data points of the control curve in *each* graph of your report. This makes it easier for the reader to grasp and compare the graphs.

Test Yourself!

These three figures present the same data set. Which of the figures is the best presentation of the data? What is wrong with the other two? (Answers are at the end of this appendix.)

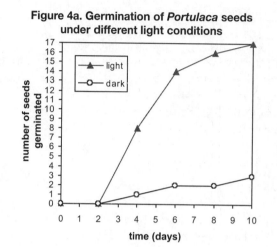

Figure 4a. Germination of *Portulaca* seeds under different light conditions

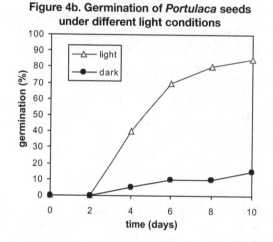

Figure 4b. Germination of *Portulaca* seeds under different light conditions

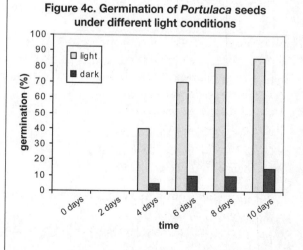

Figure 4c. Germination of *Portulaca* seeds under different light conditions

235

III. RATE DETERMINATION

Rate refers to some measurement over *time*. For example:

- What is the speed (rate of movement) of that car, in miles/hour?
- What is the growth rate of a seedling's stem, in cm/day?

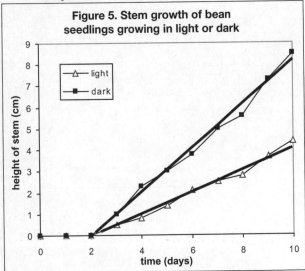

Figure 5. Stem growth of bean seedlings growing in light or dark

How to Determine the Rate by Hand

1. Plot the measurements over time as a line graph.
2. Draw a line that has the best fit to the curve.
3. Determine the **slope** of the best-fit line; the slope is the **rate**.

How to Determine the Slope of a Line by Hand

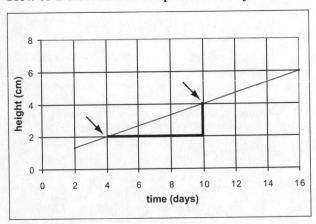

1. Select any two points on this line.
   Example: *4 cm at 10 days; 2 cm at 4 days*
2. Rise = their difference on the y-axis
   Example: *4 cm – 2 cm = 2 cm*
3. Run = their difference on the x-axis
   Example: *10 days – 4 days = 6 days*
4. **Slope** = "rise over run" = rise/run
   Example: *2 cm/6 days = 0.33 cm/day* = **rate**

How to Determine Rate Using Microsoft® Excel®

From the raw data, Excel® can automatically determine the best-fit line and calculate its slope (the rate).

1. In Excel®, type a table of the raw data (shown graphed in Figure 5):

days	height (cm) of seedlings in the light	height (cm) of seedlings in the dark
0	0	0
1	0	0
2	0	0
3	0.5	1
4	0.8	2.3
5	1.4	3
6	2.1	3.8
7	2.5	5
8	2.8	5.6
9	3.7	7.3
10	4.4	8.5

2. Click on any empty cell.
3. In the formula box in the toolbar at the top of the screen, type the equal sign and then select "SLOPE" from the function drop-down list.
4. Follow the prompts and specify the range of your data for which Excel® should determine a best-fit line (called "linear regression line").
   - The range of your data for seedlings in the *light* would be 3 to 10 (days) for the x-axis and 0.5 to 4.4 (cm) for the y-axis.
   - The range of your data for seedlings in the *dark* would be 3 to 10 (days) for the x-axis and 1 to 8.5 (cm) for the y-axis.
5. Excel® also will then automatically calculate the slope (growth rate), which appears instantly in the cell that was clicked. For this raw data:
   - The growth rate of seedlings in the light is 0.55 cm/day.
   - The growth rate of seedlings in the dark is 1.03 cm/day.

236

Figure 4a:

- A brief glance will mislead the viewer into thinking that one of the curves shows complete germination because it goes to the top of the graph area.
- A closer examination of the graph reveals only that the seeds in the light had a much faster germination rate than seeds in the dark. Nowhere is there an indication as to the percentage of germination, because the total number of seeds used in the experiment is not given.
- The vertical axis is marked off too densely, making it hard to read. For greater clarity, mark only the even numbers or mark at intervals of five.
- Using the dark symbol for "light" and the open symbol for "dark" is counterintuitive. Select symbols that are logical to the reader.

Figure 4b: Good!

Figure 4c: Do not use a bar graph for a data set that has numerically related values for the independent variable. "Time" is a numerical value that can be plotted on a linear scale; thus, use a line graph as in Figure 4b.

Appendix 2

STATISTICS

APPENDIX 3

I. Statistics—the basics
II. The t-test (by hand and by Microsoft® Excel®)
     Two-sample, two-tailed t-test
     Paired-sample, two-tailed t-test
III. Chi-square test for genetics problems (by hand)

I. STATISTICS—THE BASICS

TERMINOLOGY DEFINED

- **Population** = all items of a certain type that could be measured if it were possible to do so.
  Example: population of all twigs on an oak tree

- **Random sample** = a group of items randomly selected from a population.
  Example: 20 randomly selected twigs on an oak tree

- **Sample data** = the measurements or counts taken of the items in the random sample.
  Example: twig length that grew during 2005, measured on each of the 20 twigs in the sample

SCIENTIFIC INVESTIGATIONS

Scientists investigate a **population** by taking and measuring a **random sample** of it. To compare two populations, they take a random sample of each and compare the measurements. However, in science it is not enough to simply present superimposed histograms of the samples (as in Figure 3 of Appendix 2). A statistical test also needs to be performed on the two samples in order to determine whether there is a *significant difference* between them.

STATISTICAL TERMINOLOGY DEFINED

- **Normal curve** (or **distribution** or **population**) = a bell-shaped curve that is not too peaked and not too broad in the middle; there is a mathematical equation that defines it precisely.
- **Population mean** = μ = the average value of all the items in the population.
- **Sample size** = n = the number of data in the sample ($n=20$ if there were 20 twig lengths measured)
- **Sample mean** = $\bar{X}$ = the average value of the sample data.
- **Sample variance** = a calculated value that indicates the clustering of data points around the mean.
  - The smaller the variance, the more clustered the data points are around the mean.
  - Variance calculation is based on how far each measurement in the sample is away from the sample mean.
  - Sample variance = $s^2 = \dfrac{\sum (X_i - \bar{X})^2}{n-1}$
- **Standard deviation (SD)**
  - The sample's standard deviation = s = the square root of the sample's variance.
  - Note that in a data table, a sample mean is typically presented along with its standard deviation:
    mean ±1 SD (the mean plus or minus one standard deviation) = the range around the mean that includes 68.27% of the measurements in a normal population.
    mean ±2 SD = the range around the mean that includes 95.44% of the measurements in a normal population.

NOTE: All data presented in Appendix 3 are fictitious and for illustration purposes only.

II. THE T-TEST (by hand and by Microsoft® Excel®)
(For greater detail, see Zar 1999.)

- Use the *two-sample t-test* (see Parts II.A. and II.B.) when each datum in one sample is not associated in any way with any particular datum in the other sample; this test assumes that the samples are *random*, from 2 *normal* populations having *equal variances*. (See the "Important Notes" below.)
- Use the *paired-sample t-test* (see Parts II.C. and II.D.) when each datum in one sample is in some way associated with (paired with) one and only one particular datum in the other sample; this test does not assume normality and equal variances of the sampled populations. This test *does* assume that the subjects are randomly selected, and that the differences of the pairs in the population are normally distributed.
- For the final step in a t-test, you *must* know if the hypothesis you are testing is "two-tailed" or "one-tailed."

> *IMPORTANT NOTES:*
> - The **t-test is "robust"**: It still works if there is considerable departure from the assumptions, especially if:
>   - The two sample sizes are nearly equal.
>   - The hypothesis being tested is two-tailed.
>   - The sample size is large (Zar 1999, p. 127).
> - **Variations** on the t-test formula have been developed for use with populations having *un*equal variances. The Microsoft® Excel® "Data Analysis" menu includes a two-sample t-test assuming equal variances as well as a two-sample t-test assuming unequal variances.
> - **Nonparametric tests** exist for use with populations that deviate severely from the t-test assumptions.

TWO-TAILED VERSUS ONE-TAILED HYPOTHESES

As you will see later, the t-statistic is calculated the same way for two-tailed and one-tailed hypotheses, but the *t-critical value* (derived from Table A3.1 and compared to the calculated t-statistic) is different for two-tailed and one-tailed hypotheses. Thus, you must know whether you are investigating two-tailed or one-tailed hypotheses.

Investigation with Two-Tailed Hypotheses

The question: Was twig growth in 2002 *different* from (*either* better *or* worse than) twig growth in 2001?

Hypothesis = H_0 = the two population means are equal ($\mu_1 = \mu_2$).

Alternative hypothesis = H_A = the two population means are *different* (*either* $\mu_1 < \mu_2$ or $\mu_1 > \mu_2$).

- *Two-tailed* hypotheses: Investigate whether one population differs in *either direction* from the other population.

Investigation with One-Tailed Hypotheses

The question: A tree received fertilizer in spring 2002; was twig growth in 2002 *better* than in 2001?

Hypothesis = H_0 = the mean growth in 2002 is not better; the mean growth in 2001 is greater than or equal to the mean in 2002 ($\mu_1 \geq \mu_2$).

Alternative hypothesis = H_A = the mean twig growth in 2002 is *better*; the mean of 2001 is less ($\mu_1 < \mu_2$).

- *One-tailed* hypotheses: Investigate whether one population differs in *one direction* from the other population.

OVERVIEW OF THE TWO-TAILED T-TEST

1. The t-test uses the data of both samples to calculate a t-statistic value.
2. The greater the difference between the two samples, the larger the calculated **t-statistic** value will be.
3. The larger the t-statistic value calculated from the two *samples*, the less likely it will be that the two sampled *populations* have identical means.

4. *How large* can a t-statistic be (i.e., how different can the two *samples* be) before you reject the hypothesis H_0 that the two *population* means are the same? To determine this:

    a. Compare the calculated t-statistic value against the two-tailed critical value selected in Table A3.1 "Critical Values of the t Distribution." The critical value selected in the table is usually at the **5% level of significance** (alpha = α = 0.05).

    b. If the calculated t-statistic value is **equal to or larger** than the selected critical value in the table, then the *chance of getting two such samples when in fact the two populations have identical means* is equal to or less than 5%; therefore, **reject H_0**, the hypothesis that the two population means are the same.

    If the calculated t-statistic value is **smaller** than the selected critical value in the table, then **accept H_0**.

II.A. TWO-SAMPLE, TWO-TAILED T-TEST PERFORMED BY HAND

Example: The 2005 growth was measured on a sample of oak twigs and a sample of maple twigs. The sample data are used to test these two-tailed hypotheses about the populations:

Hypothesis = H_0 = population mean of 2005 twig lengths on oak *equals* the population mean of 2005 twig lengths on maple ($\mu_1 = \mu_2$; also written as $\mu_1 - \mu_2 = 0$).

Alternative hypothesis = H_A = population mean of 2005 twig lengths on oak is not equal to that on maple.

1. For each sample, calculate:
- Sample size = n = the number of measurements that you took in that sample
- Degrees of freedom (df) = ν (Greek letter nu) = $n - 1$
- Sample mean = $\overline{X}$ = add all the measurements and divide by sample size = $\dfrac{\sum X_i}{n}$

- Sample sum of squares = SS = for each measurement (X_i), subtract it from the mean, then square this difference, and then add all these squared differences together:
$$= \sum (X_i - \overline{X})^2$$

2. Now calculate:
- Pooled variance = s_p^2 = SS from sample 1 plus SS from sample 2. Then divide this total by the sum of the degrees of freedom of the two samples:

$$= \frac{SS_1 + SS_2}{\nu_1 + \nu_2}$$

- t-statistic = $\dfrac{\overline{X}_1 - \overline{X}_2}{\sqrt{\dfrac{s_p^2}{n_1} + \dfrac{s_p^2}{n_2}}}$

- Pooled degrees of freedom = $\nu_1 + \nu_2 = (n_1 - 1) + (n_2 - 1) = n_1 + n_2 - 2$

3. If the calculated t-statistic is a negative number, use its absolute value (i.e., ignore its negative sign).
4. Now compare the absolute value of this t-statistic to the tabled t-critical value for your pooled degrees of freedom and $\alpha = 0.05$, two-tailed [$\alpha(2) = 0.05$ in Table A3.1].
5. If the calculated t-statistic is *greater than* the tabled t-critical value, then *reject* the hypothesis that these two samples came from populations with *identical* means; "there *is* a significant difference at the 0.05 level."
6. If the calculated t-statistic is *less than* the tabled t-critical value, then *accept* the hypothesis; "there is *no* significant difference at the 0.05 level."

    NOTE: For *one-tailed* hypotheses, calculate the t-statistic in the same way, but use the t-critical value for $\alpha(1) = 0.05$ in Table A3.1. As usual, if your calculated t-statistic is greater than the tabled t-critical value, reject H_0.

ν	α(2): 0.50 / α(1): 0.25	0.20 / 0.10	0.10 / 0.05	0.05 / 0.025	0.02 / 0.01	0.01 / 0.005	0.005 / 0.0025	0.002 / 0.001	0.001 / 0.0005
1	1.000	3.078	6.314	12.706	31.821	63.657	127.321	318.309	636.619
2	0.816	1.886	2.920	4.303	6.965	9.925	14.089	22.327	31.599
3	0.765	1.638	2.353	3.182	4.541	5.841	7.453	10.215	12.924
4	0.741	1.533	2.132	2.776	3.747	4.604	5.598	7.173	8.610
5	0.727	1.476	2.015	2.571	3.365	4.032	4.773	5.893	6.869
6	0.718	1.440	1.943	2.447	3.143	3.707	4.317	5.208	5.959
7	0.711	1.415	1.895	2.365	2.998	3.499	4.029	4.785	5.408
8	0.706	1.397	1.860	2.306	2.896	3.355	3.833	4.501	5.041
9	0.703	1.383	1.833	2.262	2.821	3.250	3.690	4.297	4.781
10	0.700	1.372	1.812	2.228	2.764	3.169	3.581	4.144	4.587
11	0.697	1.363	1.796	2.201	2.718	3.106	3.497	4.025	4.437
12	0.695	1.356	1.782	2.179	2.681	3.055	3.428	3.930	4.318
13	0.694	1.350	1.771	2.160	2.650	3.012	3.372	3.852	4.221
14	0.692	1.345	1.761	2.145	2.624	2.977	3.326	3.787	4.140
15	0.691	1.341	1.753	2.131	2.602	2.947	3.286	3.733	4.073
16	0.690	1.337	1.746	2.120	2.583	2.921	3.252	3.686	4.015
17	0.689	1.333	1.740	2.110	2.567	2.898	3.222	3.646	3.965
18	0.688	1.330	1.734	2.101	2.552	2.878	3.197	3.610	3.922
19	0.688	1.328	1.729	2.093	2.539	2.861	3.174	3.579	3.883
20	0.687	1.325	1.725	2.086	2.528	2.845	3.153	3.552	3.850
21	0.686	1.323	1.721	2.080	2.518	2.831	3.135	3.527	3.819
22	0.686	1.321	1.717	2.074	2.508	2.819	3.119	3.505	3.792
23	0.685	1.319	1.714	2.069	2.500	2.807	3.104	3.485	3.768
24	0.685	1.318	1.711	2.064	2.492	2.797	3.091	3.467	3.745
25	0.684	1.316	1.708	2.060	2.485	2.787	3.078	3.450	3.725
26	0.684	1.315	1.706	2.056	2.479	2.779	3.067	3.435	3.707
27	0.684	1.314	1.703	2.052	2.473	2.771	3.057	3.421	3.690
28	0.683	1.313	1.701	2.048	2.467	2.763	3.047	3.408	3.674
29	0.683	1.311	1.699	2.045	2.462	2.756	3.038	3.396	3.659
30	0.683	1.310	1.697	2.042	2.457	2.750	3.030	3.385	3.646
31	0.682	1.309	1.696	2.040	2.453	2.744	3.022	3.375	3.633
32	0.682	1.309	1.694	2.037	2.449	2.738	3.015	3.365	3.622
33	0.682	1.308	1.692	2.035	2.445	2.733	3.008	3.356	3.611
34	0.682	1.307	1.691	2.032	2.441	2.728	3.002	3.348	3.601
35	0.682	1.306	1.690	2.030	2.438	2.724	2.996	3.340	3.591
36	0.681	1.306	1.688	2.028	2.434	2.719	2.990	3.333	3.582
37	0.681	1.305	1.687	2.026	2.431	2.715	2.985	3.326	3.574
38	0.681	1.304	1.686	2.024	2.429	2.712	2.980	3.319	3.566
39	0.681	1.304	1.685	2.023	2.426	2.708	2.976	3.313	3.558
40	0.681	1.303	1.684	2.021	2.423	2.704	2.971	3.307	3.551
41	0.681	1.303	1.683	2.020	2.421	2.701	2.967	3.301	3.544
42	0.680	1.302	1.682	2.018	2.418	2.698	2.963	3.296	3.538
43	0.680	1.302	1.681	2.017	2.416	2.695	2.959	3.291	3.532
44	0.680	1.301	1.680	2.015	2.414	2.692	2.956	3.286	3.526
45	0.680	1.301	1.679	2.014	2.412	2.690	2.952	3.281	3.520
46	0.680	1.300	1.679	2.013	2.410	2.687	2.949	3.277	3.515
47	0.680	1.300	1.678	2.012	2.408	2.685	2.946	3.273	3.510
48	0.680	1.299	1.677	2.011	2.407	2.682	2.943	3.269	3.505
49	0.680	1.299	1.677	2.010	2.405	2.680	2.940	3.265	3.500
50	0.679	1.299	1.676	2.009	2.403	2.678	2.937	3.261	3.496
52	0.679	1.298	1.675	2.007	2.400	2.674	2.932	3.255	3.488
54	0.679	1.297	1.674	2.005	2.397	2.670	2.927	3.248	3.480
56	0.679	1.297	1.673	2.003	2.395	2.667	2.923	3.242	3.473
58	0.679	1.296	1.672	2.002	2.392	2.663	2.918	3.237	3.466
60	0.679	1.296	1.671	2.000	2.390	2.660	2.915	3.232	3.460
62	0.678	1.295	1.670	1.999	2.388	2.657	2.911	3.227	3.454
64	0.678	1.295	1.669	1.998	2.386	2.655	2.908	3.223	3.449
66	0.678	1.295	1.668	1.997	2.384	2.652	2.904	3.218	3.444
68	0.678	1.294	1.668	1.995	2.382	2.650	2.902	3.214	3.439
70	0.678	1.294	1.667	1.994	2.381	2.648	2.899	3.211	3.435
72	0.678	1.293	1.666	1.993	2.379	2.646	2.896	3.207	3.431
74	0.678	1.293	1.666	1.993	2.378	2.644	2.894	3.204	3.427
76	0.678	1.293	1.665	1.992	2.376	2.642	2.891	3.201	3.423
78	0.678	1.292	1.665	1.991	2.375	2.640	2.889	3.198	3.420
80	0.678	1.292	1.664	1.990	2.374	2.639	2.887	3.195	3.416
82	0.677	1.292	1.664	1.989	2.373	2.637	2.885	3.193	3.413
84	0.677	1.292	1.663	1.989	2.372	2.636	2.883	3.190	3.410
86	0.677	1.291	1.663	1.988	2.370	2.634	2.881	3.188	3.407
88	0.677	1.291	1.662	1.987	2.369	2.633	2.880	3.185	3.405
90	0.677	1.291	1.662	1.987	2.368	2.632	2.878	3.183	3.402
92	0.677	1.291	1.662	1.986	2.368	2.630	2.876	3.181	3.399
94	0.677	1.291	1.661	1.986	2.367	2.629	2.875	3.179	3.397
96	0.677	1.290	1.661	1.985	2.366	2.628	2.873	3.177	3.395
98	0.677	1.290	1.661	1.984	2.365	2.627	2.872	3.175	3.393
100	0.677	1.290	1.660	1.984	2.364	2.626	2.871	3.174	3.390
105	0.677	1.290	1.659	1.983	2.362	2.623	2.868	3.170	3.386
110	0.677	1.289	1.659	1.982	2.361	2.621	2.865	3.166	3.381
115	0.677	1.289	1.658	1.981	2.359	2.619	2.862	3.163	3.377
120	0.677	1.289	1.658	1.980	2.358	2.617	2.860	3.160	3.373
125	0.676	1.288	1.657	1.979	2.357	2.616	2.858	3.157	3.370
130	0.676	1.288	1.657	1.978	2.355	2.614	2.856	3.154	3.367
135	0.676	1.288	1.656	1.978	2.354	2.613	2.854	3.152	3.364
140	0.676	1.288	1.656	1.977	2.353	2.611	2.852	3.149	3.361
145	0.676	1.287	1.655	1.976	2.352	2.610	2.851	3.147	3.359
150	0.676	1.287	1.655	1.976	2.351	2.609	2.849	3.145	3.357
160	0.676	1.287	1.654	1.975	2.350	2.607	2.846	3.142	3.352
170	0.676	1.287	1.654	1.974	2.348	2.605	2.844	3.139	3.349
180	0.676	1.286	1.653	1.973	2.347	2.603	2.842	3.136	3.345
190	0.676	1.286	1.653	1.973	2.346	2.602	2.840	3.134	3.342
200	0.676	1.286	1.653	1.972	2.345	2.601	2.839	3.131	3.340
250	0.675	1.285	1.651	1.969	2.341	2.596	2.832	3.123	3.330
300	0.675	1.284	1.650	1.968	2.339	2.592	2.828	3.118	3.323
350	0.675	1.284	1.649	1.967	2.337	2.590	2.825	3.114	3.319
400	0.675	1.284	1.649	1.966	2.336	2.588	2.823	3.111	3.315
450	0.675	1.283	1.648	1.965	2.335	2.587	2.821	3.108	3.312
500	0.675	1.283	1.648	1.965	2.334	2.586	2.820	3.107	3.310
600	0.675	1.283	1.647	1.964	2.333	2.584	2.817	3.104	3.307
700	0.675	1.283	1.647	1.963	2.332	2.583	2.816	3.102	3.304
800	0.675	1.283	1.647	1.963	2.331	2.582	2.815	3.100	3.303
900	0.675	1.282	1.647	1.963	2.330	2.581	2.814	3.099	3.301
1000	0.675	1.282	1.646	1.962	2.330	2.581	2.813	3.098	3.300
∞	0.6745	1.2816	1.6449	1.9600	2.3263	2.5758	2.8070	3.0902	3.2905

Table A3.1 Critical Values of the t Distribution. ZAR, JERROLD H., BIOSTATISTICAL ANALYSIS, 4th Edition, ©1999. Reprinted by permission of Pearson Education, Inc., Upper Saddle River, NJ.

II.B. TWO-SAMPLE, TWO-TAILED T-TEST PERFORMED BY EXCEL® SOFTWARE

Example: The 2005 growth was measured on a sample of oak twigs and a sample of maple twigs. The sample data are used to test these two-tailed hypotheses about the populations:

Hypothesis = H_0 = population mean of 2005 twig lengths on oak *equals* the population mean of 2005 twig lengths on maple ($\mu_1 = \mu_2$; also written as $\mu_1 - \mu_2 = 0$).

Alternative hypothesis = H_A = population mean of 2005 twig lengths on oak is not equal to that on maple.

In Microsoft® Excel®, type in your columns of data. Then in the Excel® "Tools" menu, click on "Data Analysis" to see a list of statistical tests. (If Data Analysis does not appear on the Excel® Tools menu on your computer, click on "Add-Ins" on the Tools menu instead and install "Analysis ToolPak"; after installation, you should find Data Analysis listed on the Tools menu.) From the list of tests, select "t-Test: Two-Sample Assuming Equal Variances." Fill in the required information, using the Help button if needed.

Example Printout from Excel®

This t-test was performed on a sample of 20 twig lengths from an oak and 20 from a maple:

t-Test: Two-Sample Assuming Equal Variances

	Oak	*Maple*
Mean	4.885	3.895
Variance	0.316079	0.533132
Observations	20	20
Pooled Variance	0.424605	
Hypothesized Mean Diff.	0	
df	38	
t Stat	4.804437	
P(T<=t) one-tail	1.22E-05	
t Critical one-tail	1.685953	
P(T<=t) two-tail	2.44E-05	
t Critical two-tail	2.024394	

> Pooled degrees of freedom

> Take the "absolute value" of the t-statistic. (This means ignore a minus sign if present.)

> There is a 2.44×10^{-5} chance of getting a t-stat absolute value of 4.804437 or larger when hypothesis H_0 is true.

Conclusion:
- Absolute value of t-statistic (4.804) *is greater than* the t-critical two-tailed value (2.024).
- Thus, *reject* the hypothesis that these two samples come from populations with identical means.
- The chance that the two *sample* means are as different as these are, when in fact the two *population* means are identical is *less than* 5% (the chance here is 2.44×10^{-5} or 0.0000244 or 0.00244%).
- Conclude: "There *is* a significant difference at the 0.05 level."

In other cases:
- If the absolute value of the t-statistic had been *less than* the t-critical two-tailed value, then the hypothesis that these two samples come from populations with identical means would be *accepted*.
- In such a case, there would be a *better than 5% chance* of getting such sample means when in fact the two population means are identical.
- Conclude: "There is *no* significant difference at the 0.05 level."

> *NOTE*: If the two populations have *unequal* variance, Excel® has a "t-Test: Two-Sample Assuming Unequal Variances" available that uses a somewhat different t-critical value and a different calculation for degrees of freedom.

Appendix 3

II.C. PAIRED-SAMPLE, TWO-TAILED T-TEST PERFORMED BY HAND

Use the *paired-sample t-test* when each datum in one sample is in some way associated with (paired with) one and only one particular datum in the other sample.

Example: In a random sample of 20 twigs, two measurements were taken on each twig—the length of 2001 growth and the length of 2002 growth. The sample data are used to test these two-tailed hypotheses about the populations:

Hypothesis = H_0 = in the twig population, there is no difference between twig growth in 2001 and 2002 ($\mu_1 - \mu_2 = 0$).

Alternative hypothesis = H_A = in the twig population, there is a difference between twig growth in 2001 and 2002 ($\mu_1 - \mu_2 \neq 0$).

1. Calculate:
   - The difference between the two measurements of each pair:
     $$d = X_1 - X_2$$
   - Number of pairs of data = the number of differences = n
   - Mean of the differences = average of the calculated d values = $\bar{d}$
   - Degrees of freedom (df) = ν (Greek letter nu) = $n - 1$

   - Sum of squares = SS = for each calculated difference (d_i), subtract it from the mean of the differences ($\bar{d}$); then square this amount, and then add all these squared amounts together
     $$SS = \sum (d_i - \bar{d})^2$$

   - t-statistic $= \dfrac{\bar{d}}{\sqrt{\dfrac{SS}{n(n-1)}}}$

Table 1. Twig length (cm)		
2001	**2002**	**$d = X_1 - X_2$**
5	4	1
5	4	1
5	3.8	1.2
5	4.6	0.4
4	4.2	-0.2
4.5	2.5	2
4.6	3.4	1.2
5.7	3.5	2.2
5.1	2.9	2.2
5.3	2.7	2.6
3.9	4.2	-0.3
5.9	4.1	1.8
4.9	4.8	0.1
4.3	5	-0.7
4	3.9	0.1
5.5	4.1	1.4
5	4	1
5.5	4	1.5
4.5	5.2	-0.7
5	3	2

2. If the calculated t-statistic is a negative number, use its absolute value (i.e., ignore its negative sign).
3. Now compare the absolute value of this t-statistic to the tabled t-critical value for your degrees of freedom and $\alpha = 0.05$, two-tailed [$\alpha(2) = 0.05$ in Table A3.1].
5. If the calculated t-statistic is *greater than* the tabled t-critical value, then *reject* the hypothesis that these two samples came from populations with *identical* means. "There *is* a significant difference at the 0.05 level."
6. If the calculated t-statistic is *less than* the tabled t-critical value, then *accept* the hypothesis. "There is *no* significant difference at the 0.05 level."

> *NOTE*: For *one-tailed* hypothesis where the hypothesis H_0 is that the mean population difference is less than or equal to the number **a**, and the H_A is that the mean difference is greater than **a**: the paired-sample t-statistic is calculated with the numerator being ($\bar{d}$ - **a**), and the t-critical value used is for $\alpha(1) = 0.05$ in Table A3.1. As usual, if your calculated t-statistic is greater than the tabled t-critical value, reject H_0.

II.D. PAIRED-SAMPLE, TWO-TAILED T-TEST PERFORMED BY EXCEL® SOFTWARE

Example: In a random sample of 20 twigs, two measurements were taken on each twig—the length of 2001 growth and the length of 2002 growth. The sample data are used to test these two-tailed hypotheses about the populations:

Hypothesis = H_0 = in the twig population, there is no difference between twig growth in 2001 and 2002 ($\mu_1 - \mu_2 = 0$).

Alternative hypothesis = H_A = in the twig population, there is a difference between twig growth in 2001 and 2002 ($\mu_1 - \mu_2 \neq 0$).

In Microsoft® Excel®, type in your columns of data. Then in the Excel® "Tools" menu, click on "Data Analysis" to see a list of statistical tests. (If Data Analysis does not appear on the Excel® Tools menu on your computer, click on "Add-Ins" on the Tools menu instead and install "Analysis ToolPak"; after installation, you should find Data Analysis listed on the Tools menu.) From the list of tests, select "t-Test: Paired Two Sample for Means." Fill in the required information, using the Help button if needed.

Example Printout from Excel®

This t-test was performed on a sample of 20 twigs on which both the 2001 and the 2002 growth were measured. (Note that this example uses the same fictitious data set as was used for Part II.B.; this allows you to compare the outputs of these two kinds of t-tests. This data set is shown in Part II.C. and was also used in Appendix 2 for the histogram.)

t-Test: Paired Two Sample for Means

	2001	*2002*
Mean	4.885	3.895
Variance	0.316078947	0.53313158
Observations	20	20
Pearson Correlation	-0.21302558	
Hypothesized Mean Diff.	0	
df	19	
t Stat	4.374998348	
P(T<=t) one-tail	0.000162863	
t Critical one-tail	1.729132792	
P(T<=t) two-tail	0.000325727	
t Critical two-tail	2.09302405	

Degrees of freedom

Take the "absolute value" of the t-statistic. (This means ignore a minus sign if present.)

There is a 0.0003 chance of getting a t-stat absolute value of 4.375 or larger when hypothesis H_0 is true.

Conclusion:

- Absolute value of t-statistic (4.375) *is greater than* the t-critical two-tailed value (2.093).
- Thus, *reject* the hypothesis that these two samples come from populations with identical means.
- The chance that the two *sample* means are as different as these are, when in fact the two *population* means are identical is less than 5% (the chance here is 0.0003, or 0.03%).
- Conclude: "There *is* a significant difference at the 0.05 level."

In other cases:

- If the absolute value of the t-statistic had been *less than* the t-critical two-tailed value, then the hypothesis that these two samples come from populations with identical means would be *accepted*.
- In such a case, there would be a *better than 5% chance* of getting such sample means when in fact the two population means are identical.
- Conclude: "There is *no* significant difference at the 0.05 level."

III. CHI-SQUARE TEST FOR GENETICS PROBLEMS (by hand)

The chi-square test can be used to determine whether there is a significant difference between the <u>observed</u> phenotypic proportions in the progeny generation and the <u>expected</u> phenotypic proportions calculated for that particular Mendelian genetics cross.

Hypothesis = H_0 = the observed phenotypic proportions are as expected from this particular genetic cross.

Alternative hypothesis = H_A = the observed phenotypic proportions are different from the phenotypic proportions expected from this particular genetic cross.

OVERVIEW OF THE CHI-SQUARE TEST

1. Take data: number of offspring observed in each phenotypic class.
2. Calculate: expected number of offspring in each phenotypic class for that particular Mendelian cross.
3. Calculate the chi-square statistic: For each phenotypic class, calculate the indicated fraction using the observed and expected numbers for that class; then add all these calculated values together. (k = number of classes)

$$\chi^2 = \sum_{i=1}^{k} \frac{(\text{obs}_i - \text{exp}_i)^2}{\text{exp}_i} = \frac{(\text{obs}_1 - \text{exp}_1)^2}{\text{exp}_1} + \frac{(\text{obs}_2 - \text{exp}_2)^2}{\text{exp}_2} + \cdots + \frac{(\text{obs}_k - \text{exp}_k)^2}{\text{exp}_k}$$

4. Compare the calculated χ^2 to the selected critical value in Table A3.2 "Critical Values of the Chi-Square Distribution" to determine whether the observed data is close enough to the expected in order to accept the hypothesis H_0 that the observed phenotypic proportions are as expected from this particular Mendelian genetic cross.

NOTES:
- Do the chi-square statistical test on the *numerical data* (never on the ratio or percent).
- Do not use chi-square when, for your particular sample size, the expected number of individuals in most phenotypic classes is less than five (Zar 1999); instead, increase your sample size so that you are able to do the chi-square test.
- Use the Yates correction version of the chi-square test when the expected number of individuals in each phenotypic class is between 5 and 10 (Stansfield 1991).
- Use the Yates correction version of the chi-square test when there is only 1 degree of freedom (Zar 1999, Stansfield 1991).

$$\chi^2 \text{ (Yates corrected)} = \sum_{i=1}^{k} \frac{(|\text{obs}_i - \text{exp}_i| - 0.5)^2}{\text{exp}_i}$$

The chi-square test can be used to analyze Gregor Mendel's data from the cross $RrYy$ x $RrYy$.

Hypothesis H_0: Based on Mendelian genetics, the cross $RrYy$ x $RrYy$ should yield progeny in the ratio of:

9 round yellow : 3 wrinkled yellow : 3 round green : 1 wrinkled green

Mendel observed 556 seeds:

315 round yellow, 101 wrinkled yellow, 108 round green, 32 wrinkled green

	Phenotype 1: round yellow seeds	Phenotype 2: wrinkled yellow seeds	Phenotype 3: round green seeds	Phenotype 4: wrinkled green seeds
Observed numbers	315	101	108	32
Expected numbers	$(9/16) \times 556$ total $= 312.75$	$(3/16) \times 556$ total $= 104.25$	$(3/16) \times 556$ total $= 104.25$	$(1/16) \times 556$ total $= 34.75$
$\dfrac{(\text{obs} - \text{exp})^2}{\text{exp}}$	$\dfrac{(315 - 312.75)^2}{312.75}$ $= 0.0162$	$\dfrac{(101 - 104.25)^2}{104.25}$ $= 0.1013$	$\dfrac{(108 - 104.25)^2}{104.25}$ $= 0.1349$	$\dfrac{(32 - 34.75)^2}{34.75}$ $= 0.2176$

χ^2 = the sum of these four values = $0.0162 + 0.1013 + 0.1349 + 0.2176 = 0.47$

NOTE: The larger the difference between the observed and expected values (the worse the fit of observed to expected), the larger the value of the calculated χ^2.

In this example: calculated $\chi^2 = 0.47$

Degrees of freedom (df) = ν (Greek letter nu) = # of phenotypic classes – 1 = 4 – 1 = 3
The value $\alpha = 0.05$ is typically used (discussed below).

Critical value for degrees of freedom $\nu = 3$ and $\alpha = 0.05$ in the chi-square table (Table A3.2) = 7.815

> *NOTE*: The greater the number of classes, the greater the number of terms that are added together to get the calculated χ^2 and thus the larger the calculated χ^2 value will be. The "degrees of freedom" used in the table of critical values corrects for this.

Do you accept or reject the hypothesis?

- In this example, the calculated χ^2 (deviation from the expected) is *smaller* than the critical value at $\alpha = 0.05$; thus, *accept* the hypothesis that the proposed Mendelian cross explains the observed progeny.
  This observed amount of deviation from the expected will occur, by chance, more than 5% of the time under the hypothesis H_0. There is "no significant difference at the 0.05 level."
- If, in another example, the calculated χ^2 is *larger* than the critical value at $\alpha = 0.05$, then *reject* the hypothesis that the proposed Mendelian cross explains the observed progeny.
  That observed amount of deviation from the expected will occur, by chance, less than 5% of the time under the hypothesis H_0. There is "a significant difference at the 0.05 level."

Appendix 3

The chi-square test can be used to analyze Gregor Mendel's data from the cross *Rr* x *Rr*.

Hypothesis H_0: Based on Mendelian genetics, the cross $Rr \times Rr$ should yield progeny in the ratio of:
3 round   :   1 wrinkled

He observed 7324 seeds:
5474 round : 1850 wrinkled

Number of phenotypic classes = 2
Degrees of freedom (df) = ν = phenotypic classes $- 1 = 2 - 1 = 1$

- Since there is only one degree of freedom, use the Yates correction chi-square test:

	Phenotype 1: round seeds	Phenotype 2: wrinkled seeds
Observed numbers	5474	1850
Expected numbers	$3/4 \times 7324 = 5493$	$1/4 \times 7324 = 1831$
obs − exp	−19	19
|obs − exp| (Absolute value means to ignore the minus sign if present.)	19	19
|obs − exp| − 0.5	18.5	18.5
$\frac{(\|obs - exp\| - 0.5)^2}{exp}$	$\frac{(18.5)^2}{5493} = \frac{342.25}{5493} = 0.0623$	$\frac{(18.5)^2}{1831} = \frac{342.25}{1831} = 0.1869$

Calculated χ^2 (Yates corrected) = sum of these two values = $0.0623 + 0.1869 = 0.2492$
$\alpha = 0.05$
Degrees of freedom = $\nu = 1$

Critical value for degrees of freedom $\nu = 1$ and $\alpha = 0.05$ in the chi-square table = 3.841

Do you accept or reject the hypothesis?
- In this example, the calculated χ^2 (deviation from the expected) is *smaller* than the critical χ^2 at $\alpha = 0.05$; thus, *accept* the hypothesis that the proposed Mendelian cross explains the observed progeny.
- If, in another example, the calculated χ^2 (deviation from the expected) is *larger* than the critical χ^2 at $\alpha = 0.05$, then *reject* the hypothesis that the proposed Mendelian cross explains the observed progeny.

ν	α: 0.999	0.995	0.99	0.975	0.95	0.90	0.75	0.50	0.25	0.10	0.05	0.025	0.01	0.005	0.001
1	0.000	0.000	0.000	0.001	0.004	0.016	0.102	0.455	1.323	2.706	3.841	5.024	6.635	7.879	10.828
2	0.002	0.010	0.020	0.051	0.103	0.211	0.575	1.386	2.773	4.605	5.991	7.378	9.210	10.597	13.816
3	0.024	0.072	0.115	0.216	0.352	0.584	1.213	2.366	4.108	6.251	7.815	9.348	11.345	12.838	16.266
4	0.091	0.207	0.297	0.484	0.711	1.064	1.923	3.357	5.385	7.779	9.488	11.143	13.277	14.860	18.467
5	0.210	0.412	0.554	0.831	1.145	1.610	2.675	4.351	6.626	9.236	11.070	12.833	15.086	16.750	20.515
6	0.381	0.676	0.872	1.237	1.635	2.204	3.455	5.348	7.841	10.645	12.592	14.449	16.812	18.548	22.458
7	0.599	0.989	1.239	1.690	2.167	2.833	4.255	6.346	9.037	12.017	14.067	16.013	18.475	20.278	24.322
8	0.857	1.344	1.646	2.180	2.733	3.490	5.071	7.344	10.219	13.362	15.507	17.535	20.090	21.955	26.124
9	1.152	1.735	2.088	2.700	3.325	4.168	5.899	8.343	11.389	14.684	16.919	19.023	21.666	23.589	27.877
10	1.479	2.156	2.558	3.247	3.940	4.865	6.737	9.342	12.549	15.987	18.307	20.483	23.209	25.188	29.588
11	1.834	2.603	3.053	3.816	4.575	5.578	7.584	10.341	13.701	17.275	19.675	21.920	24.725	26.757	31.264
12	2.214	3.074	3.571	4.404	5.226	6.304	8.438	11.340	14.845	18.549	21.026	23.337	26.217	28.300	32.909
13	2.617	3.565	4.107	5.009	5.892	7.042	9.299	12.340	15.984	19.812	22.362	24.736	27.688	29.819	34.528
14	3.041	4.075	4.660	5.629	6.571	7.790	10.165	13.339	17.117	21.064	23.685	26.119	29.141	31.319	36.123
15	3.483	4.601	5.229	6.262	7.261	8.547	11.037	14.339	18.245	22.307	24.996	27.488	30.578	32.801	37.697
16	3.942	5.142	5.812	6.908	7.962	9.312	11.912	15.338	19.369	23.542	26.296	28.845	32.000	34.267	39.252
17	4.416	5.697	6.408	7.564	8.672	10.085	12.792	16.338	20.489	24.769	27.587	30.191	33.409	35.718	40.790
18	4.905	6.265	7.015	8.231	9.390	10.865	13.675	17.338	21.605	25.989	28.869	31.526	34.805	37.156	42.312
19	5.407	6.844	7.633	8.907	10.117	11.651	14.562	18.338	22.718	27.204	30.144	32.852	36.191	38.582	43.820
20	5.921	7.434	8.260	9.591	10.851	12.443	15.452	19.337	23.828	28.412	31.410	34.170	37.566	39.997	45.315
21	6.447	8.034	8.897	10.283	11.591	13.240	16.344	20.337	24.935	29.615	32.671	35.479	38.932	41.401	46.797
22	6.983	8.643	9.542	10.982	12.338	14.041	17.240	21.337	26.039	30.813	33.924	36.781	40.289	42.796	48.268
23	7.529	9.260	10.196	11.689	13.091	14.848	18.137	22.337	27.141	32.007	35.172	38.076	41.638	44.181	49.728
24	8.085	9.886	10.856	12.401	13.848	15.659	19.037	23.337	28.241	33.196	36.415	39.364	42.980	45.559	51.179
25	8.649	10.520	11.524	13.120	14.611	16.473	19.939	24.337	29.339	34.382	37.652	40.646	44.314	46.928	52.620
26	9.222	11.160	12.198	13.844	15.379	17.292	20.843	25.336	30.435	35.563	38.885	41.923	45.642	48.290	54.052
27	9.803	11.808	12.879	14.573	16.151	18.114	21.749	26.336	31.528	36.741	40.113	43.195	46.963	49.645	55.476
28	10.391	12.461	13.565	15.308	16.928	18.939	22.657	27.336	32.620	37.916	41.337	44.461	48.278	50.993	56.892
29	10.986	13.121	14.256	16.047	17.708	19.768	23.567	28.336	33.711	39.087	42.557	45.722	49.588	52.336	58.301
30	11.588	13.787	14.953	16.791	18.493	20.599	24.478	29.336	34.800	40.256	43.773	46.979	50.892	53.672	59.703
31	12.196	14.458	15.655	17.539	19.281	21.434	25.390	30.336	35.887	41.422	44.985	48.232	52.191	55.003	61.098
32	12.811	15.134	16.362	18.291	20.072	22.271	26.304	31.336	36.973	42.585	46.194	49.480	53.486	56.328	62.487
33	13.431	15.815	17.074	19.047	20.867	23.110	27.219	32.336	38.058	43.745	47.400	50.725	54.776	57.648	63.870
34	14.057	16.501	17.789	19.806	21.664	23.952	28.136	33.336	39.141	44.903	48.602	51.966	56.061	58.964	65.247
35	14.688	17.192	18.509	20.569	22.465	24.797	29.054	34.336	40.223	46.059	49.802	53.203	57.342	60.275	66.619

Table A3.2 Critical Values of the Chi-Square Distribution. ZAR, JERROLD H., BIOSTATISTICAL ANALYSIS, 4th Edition, ©1999. Reprinted by permission of Pearson Education, Inc., Upper Saddle River, NJ.

LITERATURE CITED

Stansfield, W. D. 1991. *Schaum's Outline of Theory and Problems of Genetics*, 3rd ed. New York: McGraw-Hill.

Zar, J. H. 1999. *Biostatistical Analysis*, 4th ed. Upper Saddle River, NJ: Prentice Hall.

Appendix 3